FUTURE SKILLS

FUTURE SKILLS

THE 20 SKILLS AND COMPETENCIES EVERYONE NEEDS TO SUCCEED IN A DIGITAL WORLD

BERNARD MARR

WILEY

This edition first published 2022
Copyright © 2022 by Bernard Marr. All rights reserved.

Registered office
John Wiley & Sons, Inc., 111 River Street, Hoboken, NJ 07030, USA

John Wiley & Sons Ltd, The Atrium, Southern Gate, Chichester, West Sussex, PO19 8SQ, United Kingdom

Editorial Office
John Wiley & Sons Ltd, The Atrium, Southern Gate, Chichester, West Sussex, PO19 8SQ, United Kingdom

For details of our global editorial offices, customer services, and more information about Wiley products visit us at www.wiley.com.

Wiley also publishes its books in a variety of electronic formats and by print-on-demand. Some content that appears in standard print versions of this book may not be available in other formats.

Library of Congress Cataloging-in-Publication Data

Names: Marr, Bernard, author.
Title: Future skills : the 20 skills and competencies everyone needs to
  succeed in a digital world / Bernard Marr.
Description: Hoboken, NJ : Wiley, 2022. | Includes index.
Identifiers: LCCN 2022016664 (print) | LCCN 2022016665 (ebook) | ISBN
  9781119870401 (cloth) | ISBN 9781119870425 (adobe pdf) | ISBN
  9781119870418 (epub)
Subjects: LCSH: Technological literacy. | Workplace literacy.
Classification: LCC T65.3 .M34 2022 (print) | LCC T65.3 (ebook) | DDC
  602.3—dc23/eng/20220504
LC record available at https://lccn.loc.gov/2022016664
LC ebook record available at https://lccn.loc.gov/2022016665

Cover Design: Wiley

SKY10042642_020823

To my beautiful wife Claire and our three amazing children Sophia, James and Oliver.

CONTENTS

CONTENTS

INTRODUCTION
PREPARING FOR A NEW WORLD OF WORK

Technology is changing our world to such an extent that the majority of children in school today will do jobs that don't yet exist; a report by Dell and the Institute for the Future predicts that 85 percent of jobs that will be available in 2030 haven't been invented yet.[1] It's a staggering prediction. But how realistic is it? In my opinion, it's not nearly as outlandish as it sounds. Just think about how much has changed in the last decade, particularly when it comes to social media, automation, and artificial intelligence (AI).

And this pace of change is only going to accelerate as we further enter into a new industrial revolution, the fourth industrial revolution. Just as with the previous industrial revolutions, this will change the work that humans do, as many jobs evolve, some jobs become obsolete, and new jobs emerge.

In other words, the skillset and experience required from the workforce will be very different in the future. So when we think about the essential skills for success, we need to consider not just the jobs that exist today, but the sorts of jobs that may exist in the future, in a world in which more and more jobs (and parts of jobs) are given over to machines.

I don't say this to fill you with fear or dread. This isn't a "Robots are coming for our jobs. Be afraid" type of book. Far from it. I believe the future of work is bright.

Reshaping Work for the Better

Building on the previous industrial revolution (which was driven by advances in computing), this fourth industrial revolution (driven by automation and connected technologies) will continue to fundamentally reshape the future of work. Almost every job is going to change as more intelligent AIs and robots share work with us.

But far from detaching us from our humanity, I believe this wave of new technologies will make work *more* human, not less. What can be automated will be automated, leaving humans to do the work that we're ultimately better suited to—tasks that rely on distinctly human skills like complex decision-making, creativity, empathy and emotional intelligence, critical thinking, and communication. These are the sorts of skills where humans outperform even the most intelligent machine. This is where we excel. And it's where the future of work lies. In more human, more fulfilling work.

On the whole, then, I see this reshaping of work as a positive thing for those of us who are already in the workforce, and for our children, who will do jobs that we can't even imagine. (After all, the previous industrial revolutions have made our lives better. There's no reason to expect any different this time around.) But this wave of transformation comes at a time when employers around the world are already struggling to close skills gaps and fill vacancies.[2] The gap between the skills currently available within the workforce and the skills that businesses need in the fourth industrial revolution threatens to become a chasm. We must address this.

The people who will gracefully surf this wave of digital transformation will be those who have the right mix of essential future skills. Skills, then, will be a key differentiator of success. No big surprise there. But you might be surprised by some of the skills that will be most in demand.

Softer Skills Will Shine

When we talk about essential skills for this brave new digital world, many assume we're talking about tech skills, such as coding. In fact, thriving in the digital world isn't about having deep technical knowledge—rather, it's about understanding the technologies underpinning this fourth industrial revolution and the impact that these technologies will have on the future of work. It's about understanding the relative strengths of both people and technology, and how we can capitalize on those strengths. Therefore, future skills lean much more towards softer skills than you might think—towards skills that will enable humans to succeed in the digital world (not compete with machines for easily automated work). Of course, some jobs will require technical skills, but the majority of in-demand skills will be soft skills—basically, the things that machines can't do.

However, many of these skills are currently not (or very poorly) taught in traditional education settings. Schools place too much emphasis on traditional academic subjects like math; meanwhile, the enormous value in soft skills goes often unrecognized. Ironically, schools are teaching students to look good to an algorithm, instead of teaching them the skills needed to thrive in the future job market.

It's a shame because, unlike IQ, which is to some extent genetically determined, these essential skills for success are all skills that anyone can acquire and improve. Which is where this book comes in.

What to Expect from This Book

This book is for anyone who wants to equip themselves with the essential skills for success, whether you're just starting out in your career, well established in your field, considering a career change, grappling with changes in your industry, or just looking to future-proof an already successful career. There is something for everyone in this book, regardless of your chosen industry, job title, or education.

Each chapter of this book explores a key skill that businesses will be looking for in the future. I start each chapter by outlining the skill in question, then explore why that skill matters, and conclude with practical steps you can take to gain or improve that skill. With 20 skills to cover, this isn't an in-depth exploration of each skill, but it is a thoughtful summary, with helpful signposts and plenty of practical tips. More than anything, my goal is to give you the *motivation* for future learning, and provide you with a roadmap to explore these vital skills further, delving deeply into those areas where you want to or need to grow.

And I also hope this book fills you with optimism for the future of work, and indeed our world in general. Because, yes, there will be displacement of jobs—I can't pretend that millions won't lose jobs due to automation, because they will—but millions more jobs will emerge to replace those that have been lost. Furthermore, I firmly believe that technology will help us create a world that is more human, where we can leverage humans' amazing potential to solve the world's biggest problems (like climate change and inequality) and, ultimately, make the world a better place. These essential skills will help us all play our role in that vision.

So where to start? Most of the skills in this book are soft skills, but there's no denying that the ability to understand and work confidently alongside technology will continue to grow in importance. So let's begin our journey there, with the tech-adjacent skills for success, before we move onto the softer, more human skills.

Notes

1. Realizing 2030: A Divided Vision of the Future; Dell; https://www .delltechnologies.com/content/dam/delltechnologies/assets/perspectives/ 2030/pdf/Realizing-2030-A-Divided-Vision-of-the-Future-Summary.pdf

2. Preparing tomorrow's workforce for the Fourth Industrial Revolution; Deloitte; https://www2.deloitte.com/content/dam/Deloitte/global/Documents/ About-Deloitte/gx-preparing-tomorrow-workforce-for-4IR.pdf

CHAPTER 1
DIGITAL LITERACY

I recently read a brilliant book called *The New Breed*, in which the author, Kate Darling, argues that our relationship with robots should not come from a place of fear, but should be informed by our interactions with animals. According to Darling, instead of thinking in terms of humans versus machines, it's much more helpful to think in terms of our relationship with domesticated animals: as in, we're the master (for want of a better word), but they make our lives better and easier. It's an interesting counterpoint to the usual depressing predictions that robot overlords will replace us and take all our jobs.

I mention this because it's never been more important to approach technology with a positive attitude, because times are changing, faster than we ever could have imagined. (And I say that as a futurist who has built an entire career around technology trends. Even I am staggered by the accelerated pace of transformation.) In a world that's constantly changing, digital skills will quickly grow stale and need refreshing. Continual learning will become the norm. And in this ever-shifting landscape, a positive mindset—by which I mean excitement about the possibilities that new technologies bring, and a willingness to learn about them—is what will separate the successful from the not-so-successful. That's why, in this chapter, I aim to spark your excitement for a digital-driven future, a future in which digital literacy skills will become hard currency in the workplace.

What Is Digital Literacy?

In short, digital literacy refers to the digital skills needed to learn, work, and navigate everyday life in our increasingly digital world. It means being able to interact with technologies with ease and having confidence in your digital skills—from the basic digital skills to some more advanced capabilities. So we're talking about skills such as these:

- Being able to use digital devices, software and applications, whether in everyday life, in education settings or in the workplace

- Communicating, collaborating, and sharing information with others via digital tools

- Handling data in an appropriate, effective, and secure way

- Staying safe in the digital environment

- Keeping abreast of emerging new technologies

Why Does Digital Literacy Matter?

The digital transformation is probably the biggest transformation most of us have seen in our working lives. All of my work is now digital, from routine admin tasks, to creating and sharing content, to consulting with clients, to giving educational seminars. The digitization of work massively accelerated during the COVID pandemic, of course, but I expect it to continue to accelerate. The transformation will become more dramatic. And this transformation will apply across all sectors, even traditionally people-centric sectors like hospitality, education, and healthcare.

Confident and capable with technology— that's the goal

No industry will be left untouched by the digital transformation. And this means everyone's jobs will change, to one degree or another. Everyday tasks and communications will increasingly involve digital tools. Learning (whether workplace learning or full-time education) will increasingly harness digital tools. Intelligent machines—which could encompass robots, software, AIs, sensors, and who knows what else in future—will increasingly become part of every workplace, from factories to law firms.

Let's take AI, one of the biggest technology trends that we'll cover in this chapter, as an example. I believe AI is going to augment almost every job that humans do. Here's a cool example from my own business. I've been working with a company called Synthesia to create a digital me. Yes, by recording me in front of a green screen, they've been able to create a realistic digital Bernard that can say anything using my voice—all I have to do is type out what I want the digital Bernard to say and away he goes! This means I can turn one of my articles into a video of me without having to step in front of the camera (something that has enormous potential for growing my YouTube channel). There's even the potential to create videos in other languages with ease.

It's all possible thanks to AI. And pretty soon, a huge variety of occupations will use AI tools to get the job done more efficiently. Architects, for example, will be able to feed a design brief and specifications into AI-driven software and the AI will effortlessly come up with the most efficient designs for the architect to choose from. Or marketers will be able to generate rich content at the touch of a button. Or security guards will be able to analyze masses of security footage for suspicious activity, in real time. It's already happening. Just think of the rise of customer service chatbots—yet another example of AI at work.

This doesn't mean we all need to be retraining as software developers or become AI experts. But it does mean we all need be comfortable around technology tools, and develop the skills to work alongside them. With this in mind, I believe everyone should be asking themselves two key questions:

1. What does the digital revolution mean for my workplace and my job (or my future career prospects, for those still in education)? For many, it will mean that easily repeatable tasks become increasingly automated, as the division of labor between humans and machines shifts.

2. How will I equip myself with the skills needed to work alongside technology? I talk more about enhancing your digital literacy skills later in the chapter.

And looking beyond the workplace, there's no denying that digital technology is irreversibly integrated into our everyday lives. When did you last use a paper map while driving somewhere new? Or write a letter to someone? Or search for a business in one of those big, heavy local directories? My guess is it's been a while (or never, for many of my younger readers). Chances are you reach for a device when you want to find out something, communicate, navigate an unfamiliar city, or whatever. Even these everyday, familiar tasks will change rapidly as AI (and other technologies) evolve. It's therefore vital that we build a society that's comfortable, confident, and capable with technology if we want to thrive. And that requires some investment—both at a government and organizational level to equip people with the skills for success, and at an individual level, to engage with this brave new world and commit to becoming lifelong learners.

This investment in skills can't come soon enough

There's much work to do. According to one survey, 75 percent of employees think their job will become more digitally demanding within five years, yet a fifth of businesses have no digital skills strategy in place.[1] People are at risk of falling behind, in other words, due to a lack of digital literacy.

UK think tank the Learning and Work Institute makes a more urgent case and says the UK is heading towards a "catastrophic" digital skills shortage.[2] And the picture is equally troubling on the other side of the pond, where a third of US workers lack digital skills—and this despite the fact that 82 percent of middle-skills jobs (jobs that require less than a bachelor's degree while still paying a living wage) are described as "digitally intensive."[3] Something's got to change. And a big part of the solution lies in all of us embracing essential digital literacy skills.

So What Essential Digital Literacy Skills Are We Talking About?

For me, there are two levels of skills needed. First, there are the basic skills that we all need in order to use technology effectively for everyday tasks, and then there are the next-level skills that I believe are key to thriving in the workplace. We'll get into both levels in this section but ultimately, whether I'm talking about the basics or more advanced stuff, all of these skills are about being able to use technology to solve problems, communicate with others, access and share information, make work (and life) easier, and drive success.

The digital literacy basics for everyday life and work

The UK government has an essential digital skills framework that serves as a useful definition of the foundational digital skills everyone needs to navigate 21st-century life with ease. These include things like:

- Being able to turn on a device
- Connecting a device to a safe and secure Wi-Fi network
- Searching online, and finding and using websites
- Communicating with others using email and messaging apps
- Setting up and using video calling platforms such as Skype, Zoom, and FaceTime
- Sharing documents with others
- Posting on social media
- Understanding that passwords and personal information have value and need to be kept safe
- Updating and changing passwords when necessary

This may sound basic indeed, especially if you're used to working in an office where digital tools have become integrated with most tasks. But consider this: a 2018 report found that more than 11 million people (21 percent) in the UK and 10 percent of working adults lack some or all of these basic digital skills.[4]

The framework also sets out additional essential skills for work, including:

- Understanding and complying with your employer's IT and social media policies

- Complying with security policies when working remotely (Read more about cyber-threat awareness in Chapter 4.)

- Using digital collaboration tools to meet, share information, and collaborate with colleagues (A good example is using Google Docs to collaborate on a document. Read more about collaboration in Chapter 9.)

Given the rapid digitization of work, I'd also argue that basic digital literacy now goes beyond turning on devices, using technology to communicate, and the like. So, to the above lists, I would also add the following as essential basic skills that we all need:

- Having a positive attitude to new technologies and not shying away from using new tools. This includes seeing the *value* of technology, and how it can help businesses and individuals succeed.

- Generally being tech savvy, by which I mean being aware of new technologies that are emerging and considering how those technologies might impact your job.

- Understanding the potential pitfalls of technology. A good example is the "filter bubble" phenomenon. Here, search and social media platforms serve up personalized content based on what they know about you and your previous activity online, potentially creating a limited view of the world and an environment in which fake news mushrooms. This is why I consider critical thinking to be an essential future skill (read more about this in Chapter 5).

- Creating and managing your online identity and reputation, by which I mean understanding the importance of social media activity for building your personal brand, and being able to use social media *appropriately* both in and out of work. (I talk more about your personal brand in Chapter 16.)

- Creating digital content, such as blogs, videos, tweets, or podcasts, which may increasingly involve co-creating content with AI (see Chapter 8).

Next-level digital literacy skills

Now let's get into the next-level skills. If the basics are what we need to be able to navigate everyday life and do a job competently, these next-level skills are what we need to really excel in the workplace. These are the skills that will make you more valuable, and will help to "future-proof" your career (if anything can truly be future-proofed in this age of breakneck advancement).

Bear with me here, because I'm going to delve into some technical stuff like machine learning and the "metaverse." You may be asking yourself, "Do I really need to know about this?" And the answer is yes, absolutely you do. True, you don't need to understand it to the level of a software developer, for example, but you need to have a simple grasp of how AI and other related technologies will impact life and work.

(As an aside, in the future, we probably won't even need humans to have computer programming skills, because AI will be designing software for us. Elon Musk's company OpenAI has already developed an AI called GPT-3 that can generate computer code based on someone simply describing what they want the software to do, which effectively means anyone could create their own software. GPT-3 can also write articles and other content pretty much as well as human writers, even in the style of particular writers, but that's another story!)

Grasping the Potential of Artificial Intelligence

By a mile, the number-one trend everyone needs to understand is AI and machine learning. In this book, I'll often use AI as a catchall term to encompass artificial intelligence and machine learning, but that's not strictly accurate. They're not quite the same thing.

- *AI* is the wider concept of machines being able to do tasks that we would consider "smart," meaning they require a machine to "think."

- *Machine learning* is a current application of that concept in practice. In basic terms, machine learning means giving machines access to data and letting them learn for themselves, so they can solve specific problems and complete specific tasks based on what they learn from the data.

- There's also *deep learning*, which is even more advanced and cutting edge than machine learning. Deep learning gives intelligent machines the ability to cope with huge, huge data sets, such as Twitter's firehose of tweets, and, in theory, the ability to solve any problem or complete any task that requires "thought."

Whether it's based on machine learning or more complex deep learning, AI (to use the catchall term) is essentially about *using data to make more accurate predictions and better decisions*—predictions about which factory machines are likely to break down, for example, or which customers are most likely to ditch your company's product or service in the next year, and decisions such as the most efficient transit route for goods, or which leads the sales team should focus their resources on this month. That's the core of AI and its potential: more accurate predictions and better decisions, made possible by intelligent machines.

Understanding Where AI Is Headed

There's no doubt in my mind that AI will become the most transformative technology humanity has ever developed. As Google CEO Sundar Pichai puts it, AI's impact will be even greater than that of fire or electricity. The full scale of AI's potential is difficult to wrap our heads around, but let's briefly explore some of the biggest areas in which I expect to see AI breakthroughs in the very near future.

Firstly, we'll see more blended, augmented workforces. Forget those fears of robots replacing human workers. While it's true that some jobs will change due to AI, and some will be lost altogether, the majority of workplaces will become blended environments where humans work alongside intelligent machines, and businesses can get the best out of both machine and human intelligence. In the very near future, more and more of us will find ourselves working alongside intelligent tools on an everyday basis.

Then there's better language modeling. AI allows machines and devices to understand human speech, respond to spoken requests, and even generate content. This will only become more powerful in the near future. Remember the GTP-3 AI that Elon Musk's company has developed? The company is already working on a successor, GTP-4, that will be even more powerful, in theory, giving it the potential to hold conversations that are indistinguishable from human conversation.

AI in cybersecurity is another area to watch. AI is playing a greater role in cybersecurity, by learning to recognize those behaviors that may signal nefarious intentions. I expect this to be a huge focus of AI going forward. (Read more on cybersecurity in Chapter 4.)

AI will also be the lynchpin of the metaverse—a virtual world, like the internet, that we can essentially live in. More on this mind-boggling idea is coming up later in the chapter.

We will also see advances in "low-code" and "no-code" AI. Much as you can use online drag-and-drop platforms such as Squarespace to create your own website even if you don't have any web design experience, lowcode and no-code AI will allow people to create their own AI systems using easy, plug-and-play interfaces. This will do wonders for "democratizing" AI and making it more accessible to the masses.

Finally, there's creative AI. Here, we'll increasingly see AI being used for routine creative tasks, such as coming up with headlines and photo captions for articles, or designing infographics—and even for not-so-routine creative tasks, such as writing articles and creating art. Co-creation, where human creativity is enhanced with AI tools, is another key area to watch. (More on this in Chapter 8.)

It's also important to understand that AI, and the wider digitization of our world, will also impact other technology trends, such as 3D printing (indeed, it's already possible to print pretty much anything, from houses to food), or gene editing and synthetic biology. As an example of AI's wider impact, it's now possible to run digital trials of new drugs and vaccines, thereby speeding up development time.

In this way, it's important to consider AI not just as a standalone technology trend, but as an intrinsic part of a wider technology revolution.

It's All Underpinned by Data

AI wouldn't be possible without data. It's data that allows intelligent machines to spot patterns and make predictions. Therefore, another essential digital literacy skill is being data literate. I talk more about data literacy in Chapter 2, but in very simple terms, data literacy means being able to read and use data effectively. I'm not talking about becoming a data analyst here. Rather, data literacy means being able to access, interpret, and extract useful insights from whatever data is needed to do your job well and make decisions.

But data literacy also means understanding that data underpins so many other technology trends, particularly those based on AI. It's data that allows Alexa to understand your spoken requests (the technical term for which is *natural language processing*) and reply to you in natural speech (known as *natural language generation*). Data allow machines to "see," such as in autonomous vehicles, which use cameras and sensors to understand what's going on around the vehicle and act accordingly. This ability of machines to see and interpret visual data is known as *machine vision*. Then there's *robotic process automation,* in which software robots are deployed to carry out repetitive tasks, such as scheduling appointments or processing credit card applications. There's also *quantum computing*— basically, super-fast computers that are capable of carrying out tasks that traditional computers would never be able to manage. None of this would be possible without data.

Data is also connected to 5G, in the sense that better, faster telecommunications networks will allow us to carry out more data-heavy tasks on the fly, wherever we are in the world. This in turn links to *cloud computing*, because, with data stored in the cloud and better, faster networks, we'll be able to access data stored in the cloud from anywhere. But 5G networks will also enable more *edge computing*, where data is processed on devices rather than in the cloud.

The proliferation of data has also given us smart, well, everything, from smartphones to smart homes and even smart cities. This will only continue as everything in our lives becomes smarter, from our fridges and vacuum cleaners to our workplaces.

Living in a Digital Universe: The Metaverse

Could the future of the internet be us *living in the internet* rather than just looking at it? That's the idea behind the *metaverse* concept, and it's the next big digital trend after AI. Mark Zuckerberg has said building a metaverse is something he was interested in before he ever dreamed of Facebook. But what is a metaverse? It's the term for a persistent, shared, virtual 3D world, in which more and more activities—working, gaming, going to a concert, shopping, hanging out with friends, and more—take place in a virtual, not physical, environment. "Shared" is a key word there, since the metaverse is all about creating a shared, immersive experience where people can collaborate and interact as though they were in the same physical space. The metaverse doesn't need to be limited to one platform, but there does need to be a shared, continuous experience. So you could move from an immersive Virtual Reality (VR) environment to a 2D application on your phone, but the key thing is there's continuity between the activities and environments. Having your own individual digital avatar—a digital you—that represents you across different experiences will be a key feature of the metaverse.

The idea of humans being permanently plugged into machines experiencing an immersive digital reality naturally raises unfavorable comparisons with *The Matrix*. And of course there are moral and ethical challenges to consider, such as the potential for anonymous trolls to stalk us across immersive digital spaces. But, if you think about it, the metaverse is a concept that humanity has been naturally building towards since the emergence of the internet, social media, shared digital environments such as Second Life, and virtual and augmented reality. In other words, more and more of our everyday activities are already taking place in a digital environment—something that the pandemic only accelerated—and the metaverse could be the next logical step on that journey.

If this sounds far-fetched, consider the ever-popular Fortnite game as an example. The game has begun hosting virtual concerts on the platform, attended by millions of players who can watch artists like Ariana Grande perform a set *within the game* instead of in real life. This is a sign of what's to come, if the metaverse concept comes to fruition.

Even more, let's say, vintage artists are getting in on the trend for immersive, digital experiences. Swedish supergroup Abba has worked with visual effects company Industrial Light & Magic to create digital versions of themselves in their prime—creating virtual copies of the foursome that behave accurately in every way, right down to every dance movement. And this de-aged virtual Abba (ABBAtars, as they've been dubbed) will be taking to the stage in London in 2022 in a digital performance named Abba Voyage. Fans will go to a purpose-built, physical venue in London, but they'll be treated to virtual avatars, depicting the group as they were in 1979. It's an intriguing glimpse at the future of entertainment, one that could potentially see Elvis resurrected for concerts, and who knows what else.

But realistically, how close are we to developing a metaverse? Companies like Facebook and Microsoft that are exploring this area mostly position the metaverse as an aspirational thing to aim for. So it's not like it's just around the corner. For now, the metaverse is more of a concept for making existing online environments more immersive and even more deeply integrated with our lives—for instance, by merging virtual reality with social media, something that Facebook has said it hopes to do within the next five years.

Looking further ahead, living inside the internet will be possible thanks to new, more immersive devices and hardware. For example, instead of using chunky VR headsets to enjoy a virtual experience, we'll be able to put on a pair of comfortable smart glasses—and beyond that, potentially wear smart contact lenses. The line between the real world and the digital world will become all the more blurred as we find new ways to plug into digital experiences. (By the way, if you're interested in reading more about this topic, check out my book *Extended Reality in Practice*.)

Clearly social media and virtual reality are key stepping-stones on the way to unlocking the metaverse. But, in my opinion, another important stepping-stone comes in the form of Omniverse, developed by gaming and AI pioneers Nvidia. Omniverse is a simulation and collaboration platform that runs physically realistic virtual worlds and connects to other digital platforms. At this point, it's mostly designed to coordinate remote teams and give them digital environments that recreate the experiences of working together in real, shared workspaces—including fully animated avatars created from webcam feeds. Tools like this could revolutionize the nature of work as more and more of us switch to remote working. But it's also easy to see how a platform like this could be extended to all kinds of nonwork experiences, such as having a virtual quiz night with friends.

How to Improve Your Digital Literacy

Let's explore some useful starting points for boosting your digital literacy.

Where should individuals start?

The first step is to understand where you're currently at in terms of digital literacy. Start with the basic skills outlined in this chapter and assess whether you have the knowledge needed to use technology effectively in everyday life. Depending on where you are in the world, there should be some useful government or institutional learning resources to help you pick up essential digital skills. A great example in the UK comes from the Open University, which has a free course called "Digital Skills: Succeeding in a Digital World," designed to help people develop the confidence and skills for life online.

When it comes to the next-level technology trends, such as AI and the metaverse, let me stress again that we won't all have to become tech experts to succeed. Rather, you need to be aware of these technologies, and consider how they might impact your work and life. How you keep abreast of these trends will depend on your preferred way of learning new things. There are resources like WIRED magazine, and the GeekSpeak podcast. Or there's tons of accessible information on YouTube, and a whole host of free online courses on particular topics, designed for every level from beginner to pro.

There's also my own website, bernardmarr.com, which provides a wealth of info on all manner of technology trends, plus practical case studies that show how businesses are already using these tech tools to drive success. And don't forget to check out my YouTube channel for video content, from short and accessible videos to deep dives on certain topics. Just search my name on the platform, and hit subscribe.

And, of course, do encourage your employer to invest in digital literacy training and support. This will be a harder task at some companies than others, but try to sell the positive benefits that come with enhanced digital skills—including improved productivity and performance.

Ultimately, though, the very best thing you can do for your digital literacy skills is to think of yourself as a "lifelong learner." (Read more about continual learning as an essential skill in Chapter 18.) And the second-best thing you can do is to approach new technologies with a positive mindset. Because, yes, the digital transformation will change many people's jobs, and will lead to many millions of jobs becoming obsolete. But the number of new jobs being created by the digital revolution will outnumber those lost. To put this in numbers, the World Economic Forum estimates that 85 million jobs may be displaced by 2025 as the division of labor shifts from humans to machines, but, crucially, 97 million new roles will emerge that are more suited to this new division of labor.[5]

What should employers be doing?

Employers will also need to take steps to ensure their workforce has the digital skills necessary for success. The starting point is to understand the current state of digital literacy in the organization, and identify any existing gaps and training needs. Good questions to ask include:

- How comfortable are employees using current technology tools?

- How well do they adopt new tools that are introduced?

- Do they understand how digital tools benefit their work? (Even younger employees who have grown up with technology may not necessarily value the benefits of it in the workplace.)

- Are they comfortable using social media, and do they understand how to use it to advocate for your brand?

- Do teams routinely collaborate using digital tools?

- Is everyone clear on cybersecurity risks, and do they understand how to protect themselves and the organization?

- And are they able to interact with data in an ethical and safe way?

Employee surveys (and, potentially, digital tests) will help you understand the current state of digital literacy. Only then can you put together a digital learning program and provide ongoing support—harnessing both online learning and on-the-job training as appropriate. But remember, a key part of boosting digital literacy in the workplace is emphasizing time and time again why it's valuable for every employee. After all, change can be scary, so it's important to cultivate a positive attitude to technology and dispel those negative stereotypes of robots coming for people's jobs. Couple this with an organizational culture that values lifelong learning and you're in a great position to face the rapid transformation coming our way.

Key Takeaways

Let's finish up this chapter with some final key takeaways.

- Digital literacy means having the essentials skills needed to navigate our increasingly digital world—at work, in everyday life, and in education.

- Maintaining a positive attitude is vital in this fast-changing environment. Forget doomsday predictions of robot and AI overlords. Instead, try asking, "How will technology help me do a better job, live life with greater ease, and achieve my personal and professional goals?"

- We will all have to become lifelong learners, so start now and invest some time reading up on tech trends that you're not so familiar with.

- Digital literacy and the digitization of work will not come at the expense of more human skills, such as empathy and critical thinking. If anything, it'll make our innately human capabilities all the more valuable in the workplace. Read more about these essential human skills in later chapters.

The massive digital acceleration we're experiencing wouldn't be possible without data. So let's turn to the essential data-related skills needed to thrive in the 21st-century workplace.

Notes

1. Workplace digital literacy strategies; Helastel; https://www.helastel.com/
workplace-digital-literacy-strategies-what-decision-makers-and-
employees-think/

2. UK 'heading towards digital skills shortage disaster'; BBC News; https://
www.bbc.com/news/business-56479304

3. A Third of US Workers Lack Digital Skills; BusinessWire; https://www
.businesswire.com/news/home/20210615005215/en/A-Third-of-US-
Workers-Lack-Digital-Skills

4. Essential digital skills framework; Department for Education; https://
www.gov.uk/government/publications/essential-digital-skills-framework/
essential-digital-skills-framework

5. The Future of Jobs Report 2020; World Economic Forum; https://www
.weforum.org/reports/the-future-of-jobs-report-2020/digest

CHAPTER 2
DATA LITERACY

We're living through the fourth industrial revolution (or "Industry 4.0," as it's sometimes known), a revolution that's defined by wave upon wave of new technologies that combine the physical and digital worlds. You've no doubt noticed the plethora of "smart" everyday devices, everything from watches to fridges, that are now connected to the internet. That's the fourth industrial revolution in action. And it's all underpinned by data. Data is the fuel that powers this new age of constant technological breakthrough.

As a result, data is now a prized business asset, and organizations of all kinds will want to employ data-literate individuals who can help them extract value from data. And this means everyone must understand the basics of how to use data. In fact, I'd go so far as to say data literacy is one of the most important future skills in this book—data literacy is to the 21st century what literacy was in the past century.

What Is Data Literacy?

Data literacy means a basic ability to understand and use data. That's it in a nutshell. So, in an average business context, this will generally mean being able to:

- Access appropriate data—by which I mean having access to the data needed to do your job and make informed decisions.

- Work with data—which may include creating data, gathering data, managing data to ensure it stays up to date, and of course, keeping data safe.

- Find meaning in the numbers—including understanding what the data is and what it represents, analyzing the data, and uncovering actionable business insights and opportunities.

- Communicate those insights to others in the business—being able to tell a compelling story or communicate a particular message to the right audience, based on what the data tells you, is vital for turning insights into action.

- Question the data—blindly following data is never a good idea. So an important part of data literacy is asking questions such as "Where has this data come from?" "Is this data valid?" and "Is the data biased?"

I get that a lot of people are scared of data, and I'll talk more about that later in the chapter. But love it or loathe it, there's no denying that in this fourth industrial revolution, all employees, not just data scientists, will need to acquire these must-have data literacy skills and be able to confidently work with data. So let's explore why data literacy is a such a vital skill.

Why Does Data Literacy Matter?

At the start of this chapter, I said data was an important business asset. But the truth is data is arguably *the* most important business asset. Indeed, it's now considered the most valuable resource in the world—even more valuable than oil.[1] The giants of the fourth industrial revolution are companies like Alphabet (Google's parent company), Facebook, and Amazon—all companies that are built from the ground up on data. Data also underpins a lot of what I talked about in Chapter 1. AI, for example, relies on machines being able to learn from huge amounts of data. Therefore, a basic understanding of data gives you a good foundation for learning about other technologies.

Ultimately, data is everywhere

Data isn't just exploding in importance; it's literally exploding in volume all the time. Every day, every second of every day, more and more data is created. As of 2020, there was estimated to be 44 zettabytes of data in the world, and by 2025 there could be as much as 175 zettabytes of data.[2] What on earth is a zettabyte? To put it in context, a zettabyte is 1,000 exabytes, and an exabyte is 1,000 petabytes, and each petabyte is 1,000 terabytes, and each terabyte is 1,000 gigabytes. In other words, a zettabyte has 21 zeros on the end of it! It's insane to imagine that much data existing in the world.

Or is it? Because almost everything we do these days generates data. Every single interaction with your phone or computer. Every post and like on social media. Every time you walk down the street with a phone in your pocket, sending GPS signals on your location. Every time you pass by a security camera. Every time you tap and buy something with your contactless credit card. Every time you stream a movie or listen to a podcast. It's all creating data. And because this vast stream of data is only going to get bigger, it makes sense that more and more jobs are going to involve working with data in one way or another.

Explosive demand for data skills

Even for more traditional businesses, data is fast becoming the most critical business asset. Data is what allows companies to make better decisions, understand their customers better, and streamline business operations. Think about the average marketing role, for instance. Among many other things, data gives marketers valuable insights on customer demographics, so they can run targeted campaigns. Data isn't just "an IT thing," then. It's a vital part of most modern business functions.

And this means employers increasingly need people with data skills, from the basic to the advanced. A study by the Royal Society in the UK found that demand for data scientists alone had tripled over the past five years, rising 231 percent.[3] But it seems there aren't enough people out there with those much-needed data skills, since a government report on the data skills gap found 46 percent of UK employers had struggled to recruit for data-related roles in the last two years.[4] So, on the one hand, data is the most valuable resource in the world. Yet data is also one of the main roadblocks to a company's success—largely because of a widespread lack of data skills.

The good news is that you don't need to be a data scientist to be successful. Because data isn't just for data scientists. (Although if you do fancy a career change, there's obviously a lot of demand for data scientists!)

Data touches so many roles in the average company. And this is why everyone should be able to use data to influence both their day-to-day activities and big-picture decisions. Used well, data can help you achieve your objectives at work, do a better job, and contribute to the company's overall performance. And, given the demand for these skills, and the data skills gap that exists in so many organizations, being data literate marks you out from the crowd.

Here are just a few of the ways data literacy can help you do a better job:

- You can solve problems more easily and make better decisions. Every job at every level requires some degree of problem-solving and decision-making. Whether you want to reduce waste, pinpoint key customers, increase sales, or whatever, data can help you do it.

- What's more, you don't have to rely on others for information. With modern data analytics tools, pretty much anyone can interrogate data and uncover useful insights. So if you can understand the basics of data, you can get on with more tasks yourself instead of having to wait for the analytics team to pull basic reports for you.

- Plus, you can make a more compelling case to stakeholders. Ever struggled to get buy-in for a new project or resourcing for a new initiative? Data helps you communicate your message and back up your arguments with hard evidence—meaning your pitch, whatever it's for, is more likely to be green-lit.

- You can communicate with technical colleagues more easily. These days, IT and technology functions aren't squirreled away in the company basement, doing their own thing. Cross-functional collaboration is the norm in most businesses, which means you need to be able to "speak data" with your tech-minded colleagues and ask the right questions.

So What Do You Need to Know About Data?

Remember, we're not talking about advanced data skills and knowledge here. You don't need to become a statistician or data scientist to be able to get the best out of data. What you're aiming for is proficiency and confidence when working with data. To achieve this, you'll need to understand a few key things about data. Starting with some basic data terminology.

Data speak 101

Bear in mind that I could write a whole book on the basics of data, and there's only so much I can cover in this brief summary. You'll need to do some self-study to learn more about the basics of data, and I give some recommendations for that towards the end of the chapter.

So what is data? Data is information, basically. Traditionally, data would be things like numbers and statistics. But these days, data can be any kind of information, including photos, videos, and text. Spoken commands to your Alexa device, social media updates, images you post online . . . it's all data.

Whatever the source, data always falls into one of two categories. There's *quantitative data*, which is anything that can be counted and measured (such as the price of a tub of ice cream, and how many tubs are sold). And there's *qualitative data*, which covers things like characteristics, perceptions, feelings, and descriptions (such as the flavor of ice cream, and how people feel about that flavor). In essence, quantitative data is the numbers stuff, while qualitative data is more descriptive.

Data may represent a one-off snapshot in time—such as a customer satisfaction survey—in which case, it's known as a *cross-sectional* study. Or it can be a *longitudinal* study, which means it's measured repeatedly over time to show how values change—monthly sales data being a good example.

Another term you might hear bandied around is *data set*. A data set is just a collection of data. And within the data set itself, each data point—whether ice cream sales, flavors, customer satisfaction scores, or whatever—is known as a *variable*.

Now that we've covered the basic terminology, let's look at some other things you need to know about data.

Data comes from a variety of sources

There are infinite sources of data out there to tap into. Broadly speaking, these sources fall into two categories. First, there's *internal data*, which is information gathered within your company—think sales and revenue reports, employee data, customer data, transaction records, business emails, and so on. And second, there's *external data*, which is any data collected outside the business. Some of these external data sources will be free (such as government data or Google Trends data), while others you have to pay to access (such as data from specialist providers).

Not all data is created equal

Data can be good quality or it can be bad, so you want to make sure you're working with high-quality data. (There's more on questioning and challenging data coming up later in the chapter.) In essence, good data is data that's:

- Accurate
- Consistent
- Current

- Complete (or as complete as possible). This may be as simple as understanding what "total revenue" actually includes in the context of a particular data set.

Data is meaningless without analysis

It's only when you analyze data that you can unearth interesting or valuable insights. Generally speaking, data analysis means looking for patterns and trends, and these hopefully tell a story that can inform future actions and decisions. Thanks to off-the-shelf AI-based platforms such as Amazon Web Services and IBM Watson, any business, big or small, can access intelligent tools to help them make sense of data—which means people right across the business, and not just analytics professionals, can get the best out of data. Your business may employ a range of different analytics tools, in which case part of being data literate means being able to select the most appropriate tool for the task at hand.

As I've said already, you don't need to be a data scientist to analyze data, especially with the rise of *augmented analytics* tools. With augmented analytics, data is automatically taken from data sources, analyzed, and communicated in a report using natural language processing (see Chapter 1) that nontechnical people can easily understand. To put it another way, augmented analytics looks for patterns and other valuable insights in the data, without needing data analysts to make sense of the data. This will drastically open up data analytics to a much wider array of businesses, helping to democratize data and turn all organizations into data-driven organizations.

(By the way, this doesn't mean the end of data scientists. Rather, the work of data scientists will move away from repetitive tasks that are easily given over to machines, and instead focus on more strategic and creative tasks, such as asking better business questions.)

Data should be at the heart of all decision-making

One of the things that makes data so powerful is that it can help you solve your biggest business challenges and answer your most pressing business questions. Therefore, a big part of getting the best out of data involves identifying the questions you most want answered, and then finding the best data to answer those questions. Maybe the data you need already exists in the organization, or maybe you need to gather new data with the help of others in the business. Either way, data can help you fill in the blanks, so you can make more informed decisions—both in your everyday work and for the big-picture decisions.

Let me put it this way: data for data's sake is pointless. You can have the biggest data set in the world, but if you're not using it to answer questions, solve problems, make informed decisions, and drive action, what's the point?

Data makes people nervous

Plenty of people love numbers, but there's a significant (arguably larger) portion of the population that really, really doesn't like numbers. As such, the word "data" provokes negative reactions in many people, ranging from mistrust and avoidance to outright fear and phobia. (There's even a name for it: arithmophobia, or the irrational fear of numbers.)

There are many reasons why people might not be thrilled about becoming data literate. Maybe they hated math at school (which is extremely common—one study found that six out of ten university students suffered from diagnosable math anxiety). Maybe they're worried about their job changing or becoming obsolete. Maybe they don't like asking questions for fear of looking stupid. Whatever the reason, fear stops people from trying new things, and it's the enemy of data literacy. Greater exposure will help overcome any fear of working with data, which is why it's a good idea to get used to your company's data and analytics systems sooner rather than later (more on boosting your data literacy later in the chapter).

Communicating insights from data is a valuable skill in itself

When you uncover insights within data, chances are you'll need to communicate those insights to others in the business (and potentially other stakeholders outside of the business). For example, you may want to use what the data tells you to support a new project, or make the case for more marketing spend, or propose a new product or service, or whatever. Data will help you make your case and get buy-in from others. But you must be able to present that data in an engaging, easily digestible way (especially considering how many people feel about numbers). Ultimately, the goal isn't to overwhelm people with how impressive your data is, but to ensure the data is *understood.* If the data reveals a story, your job is to find the best way to tell that story.

Data visualization is a great way to tell a story, because, as the saying goes, "A picture paints a thousand words." There are plenty of data visualization tools out there—indeed, your company's analytics tools probably include some sort of visualization element, from simple graphs to trendy infographics. When it comes to presenting data visually you should aim to:

- Use benchmarks, such as a percent change, to help people easily see the difference between two numbers.

- Use colors—for example, red for a negative percent change and green for a positive percent change.

- Use pictures or graphics to convey positive and negative changes, such as check marks, pluses and minuses, or even weather symbols like sun, grey clouds, and storm clouds.

- But don't forget about words. Everyone digests information differently, and for some, the easiest way to understand the meaning in a set of numbers is through words, not visuals. So use simple headlines and short descriptions to highlight the meaning behind the visual, or to give a little context to numbers.

Data must be challenged

I believe critical thinking is another important future skill—so much so that I've devoted a whole chapter of this book to the topic. From a data point of view, it's really important to question and challenge data, rather than simply assume data is infallible, because, no matter how comprehensive a data set is, it's never going to be 100 percent complete—which means there will always be some level of uncertainty in what the data tells us. More than that, data can often be downright biased or skewed towards certain groups. Therefore, it's always a good idea to ask questions about the data you're working with—questions like:

- Where did this data come from? Did it come from a reputable source?

- Is this the right data for the task at hand? After all, different types of data are relevant to different tasks.

- How current is the data?

- How representative is the data? Is there potential bias within the data (or indeed, from the people looking at the data)?

- What is missing from the data?

- How has the data been analyzed?

Expensive mistakes can be avoided by simply questioning data. The Enron scandal, for example, was largely down to bad data. A simple audit would have identified that the data was fraudulent and saved shareholders the loss of billions of dollars. It's an extreme example, but it shows what happens when people don't question the data in front of them.

Bias in data is a particularly hot topic to be aware of. Data bias essentially means that certain elements within the data set—genders, races, etc.—may be more heavily represented or underrepresented than others. One of the great promises of artificial intelligence was that it would eliminate bias—after all, machines don't bring the same baggage as humans—but it turns out that AI systems can be just as biased as humans, largely because of the data that these systems are trained on. By some estimates, pretty much all big data sets are biased. And this is bad because it can produce results that are discriminatory and harmful. Amazon, for example, had to shut down a program that scored candidates for employment because it was penalizing female candidates.[5] Solving the data bias problem is way beyond the scope of this book, but as a data-literate person, it's really important to be aware of the potential biases within data sets, and how these may affect your outcomes.

Of course, there are also biases in the way people respond to data. Research has shown that people can make completely different decisions based on identical information. Therefore, data-based decision-making can still be shaped by people's underlying beliefs and decision-making styles. This is why it's important to question decisions as well as the data itself. (Turn to Chapter 5 to read more about critical thinking, and Chapter 6 for more on decision-making.)

Correlation isn't the same as causation

It's a common mistake, thinking correlation and causation are the same thing. But just because there's a relationship between two variables doesn't mean one causes the other. To be clear, *correlation* means two or more factors tend to be observed at the same time. While *causation* means one directly causes the other(s), they're completely different things, yet they're often confused, especially in data science. But let me put it this way: just because there's a correlation between the divorce rate in Maine and margarine consumption—an actual real-world correlation, by the way[6]— doesn't mean Maine residents should avoid eating margarine to save their marriages!

Confusing correlation with causation can lead to poor decision-making, so always be wary of patterns in data and don't automatically assume that one variable causes another.

Data privacy and ethics are going to be increasingly important areas

Your company will already have data usage and security policies that you need to comply with. But going above and beyond compliance, data literacy also requires you to understand the ethical pitfalls around data. So much data contains personal information, and this is valuable stuff that needs to be protected and used responsibly. This will only become more important as regulators step up efforts to govern data collection and usage.

For me, good data governance means several things. For one thing, it means only gathering data that is business critical—as in, don't just gather data for the sake of it. It also means making people aware of what data you're gathering from them, why you're gathering that data, and how it'll be used. And it means giving people the chance to opt out wherever possible.

Of course, there's also a need to protect data from cyberattacks. (You can read more about cyber-threat awareness in Chapter 4.)

How to Improve Your Data Literacy

One study by Accenture highlights the stark reality of data literacy in the business world; while 75 percent of C-suite execs believe that all or most of their employees have the ability to work with data proficiently, only 21 percent of employees (across a variety of roles) were actually confident in their data literacy skills.[7] And in education, many students also feel ill-prepared for using data—one study shows that 47 percent of students find the concept of data analysis to be scary.[8]

Clearly, something is going wrong here. There needs to be widespread investment in data literacy skills, at a government level, in formal education settings, in organizations, and among all of us as individuals. While government and education are beyond the scope of this book, let's explore what individuals and employers can do to boost data literacy.

For individuals

First things first: everyone should be encouraging their employer to create a data literacy program (more on this coming up later). And in the meantime, to get comfortable using data, you can start delving into your company's data sets, using whatever management dashboards or business intelligence tools your company has. And if you don't have access to data in your role, ask for it!

There are also many self-study options online that will help you navigate data—covering everything from the basic data skills to advanced machine learning skills. A good starting point is to check out education platforms like Coursera, Udemy, and edX, as well as the excellent learning resources from the Data Literacy Project. You'll also find that there are specific data literacy courses for different industries, such as healthcare. (Coursera, for example, has a course on healthcare data literacy.)

I'd also recommend taking a basic statistics course, because this will help you understand the foundations of data and analytics, and a basic data visualization course, because this will help you communicate insights from data to others in the business.

Over and above taking an active learning approach, the best thing you can do for your data literacy is not to let fear or hesitancy around data stop you from becoming data literate. I get that there's a lot to be nervous about with data. But you really can't afford to be left behind. Data literacy will be one of the most vital skills anyone can have in the future, so try to acknowledge any fear and then find your way around it. For some people that'll mean reading whatever they can get their hands on until the topic becomes normalized. For others, it may mean simply diving in and having a go. Burying your head in the sand isn't an option.

For employers

Data literacy will be different for every business but, as a general rule, employers should look to create a baseline of foundational data literacy skills, and create a common language around data. As someone who's helped companies develop their own data literacy programs, here are my tips for boosting data literacy across your business:

Step 1 is to understand your current level of data literacy. For example, how many people are actually using data on a regular basis to make decisions? Do managers routinely use data to back up proposals for new initiatives?

Step 2 is to identify your fluent data speakers and data "translators." You may already have data analysts who are used to speaking fluently about data. But you also need "translators"—data-literate people from different business functions who can bridge the gap between the tech folk and various business groups. As part of this, also look at gaps in your data communication—meaning where are the communication barriers that prevent data from being used to its full potential?

Step 3 is to sell your people on the benefits of data literacy. If you can explain why data literacy is essential for the business's success, it'll be much easier to get people on board with data literacy training. Remember, this is an area that many people are wary of, so also try to emphasize the benefits to people's individual roles.

Step 4 is to use examples and stories to demonstrate successes. Part of selling people on the benefits of data literacy is showing how it drives business success. At first, you can use case studies from other organizations, in or outside of your own industry. (There are plenty of use cases online, including on my website, and I also have a book of case studies called *Big Data in Practice*.) Over time, you'll be able to share stories from within the organization, showing how others in the business have used data successfully.

Step 5 is to ensure access to data. It's vital that everyone is able to access, manipulate, analyze, and share the data needed to do their job well. There are plenty of management dashboard and visualization tools out there to make this easier.

Step 6 is to create a data literacy program, but start small. There's no one single way to establish a data literacy program, and you may need different learning paths for different business functions. But it's likely to involve training on business-wide tools, plus job-specific skills. Whatever you do, start with one business unit at a time and use what you learn from that pilot program to make improvements for the next group. And try to make learning fun and engaging. Data training doesn't have to be boring.

Step 7 is to lead by example. Ultimately, you want to create a data-first organizational culture, where data is prioritized at all levels of the business. To aid this, leaders need to prioritize data in their own work, for example, by using data to support decision-making.

And finally, step 8 is to build a culture of continual learning. After all, this is an area that will continue to evolve, so you want to foster an environment where continual learning and curiosity is rewarded.

Key Takeaways

Let's quickly recap some of the key points on data literacy:

- Data is the most valuable resource in the world—more valuable even than oil. Therefore, employers will increasingly want people with data skills, from the basic to the advanced.

- Data literacy simply means being able to understand and work with data with confidence. It doesn't mean becoming a data scientist (although it's a fantastic career opportunity if you're that way inclined). Rather, it means being able to access and interrogate data in your everyday job so that you can pull out valuable insights, make better decisions, and so on.

- An important part of data literacy is being able to question data and consider the potential pitfalls of data, such as data bias. Blindly following data is never a good idea.

As well as data literacy, wider technical skills will also become more valuable in the fourth industrial revolution. True, AI and automation will mean machines take on more and more tasks, but we'll still need humans with technical skills—data scientist being a great example. In the next chapter, I explore valuable 21st-century technical skills in more detail.

Notes

1. The world's most valuable resource is no longer oil, but data; *Economist*; https://www.economist.com/leaders/2017/05/06/the-worlds-most-valuable-resource-is-no-longer-oil-but-data

2. How Much Data Is Created Every Day?; Seed Scientific; https://seed scientific.com/how-much-data-is-created-every-day/

3. Royal Society: Dynamics of Data Science; Burning Glass Technologies; https://www.burning-glass.com/research-project/royal_society_dynamics_data_science_skills/

4. Quantifying the UK Data Skills Gap; Department for Digital, Culture, Media & Sport; https://www.gov.uk/government/publications/quantifying-the-uk-data-skills-gap/quantifying-the-uk-data-skills-gap-full-report

5. Understanding Data Bias; Towards Data Science; https://towardsdata science.com/survey-d4f168791e57

6. Spurious Correlations; Tyler Vigen; https://www.tylervigen.com/spurious-correlations

7. The human impact of data literacy; Accenture; https://www.accenture.com/us-en/insights/technology/human-impact-data-literacy

8. Data literacy skills crucial for the workforce of tomorrow; TechRadar; https://www.techradar.com/uk/news/for-the-workforce-of-tomorrow-data-literacy-skills-are-crucial

CHAPTER 3
TECHNICAL SKILLS

Don't skip this chapter thinking it's all about IT or engineering-type skills. It's not. "Technical skills" spans the huge variety of "hard" skills that are necessary for many jobs. With the rise of automation, we could potentially lose these vital technical skills as more tasks are given over to machines and fewer people choose to learn technical skills. And if we lose technical skills among the human workforce, we risk losing them forever within the space of a generation.

So while we all know the nature of work is changing, and workplaces will undoubtedly become more automated, there's still enormous value in technical skills. In fact, in the complex, hybrid workplaces of the future—where tasks and goals are accomplished through a blend of machine and human power—technical skills will become more valuable than ever.

What Are Technical Skills?

As you can probably tell from the previous two chapters, technical skills around coding, AI, and data science are already in high demand. That demand will continue to grow. Yet the term "technical skills" stretches far beyond technology and IT fields.

Technical skills are, in essence, the "hard" practical skills needed to do a job successfully. If you're an accountant, your professional accountancy skills are technical skills. Same goes for a plumber, nurse, truck driver, lawyer, teacher, hairstylist, project manager, carpenter, and so on. These all require knowledge and skills that are particular to that field, and you may gain those skills through a combination of training, education, on-the-job learning, and good old-fashioned experience. Even jobs that don't immediately appear to be technical or specialized often require some sort of technical skill, such as using a customer database or point-of-sale technology.

Technical skills may therefore be digital, or they may be scientific (think of a biologist, or someone working in nuclear energy), or they may also be practical and physical, requiring knowledge of specific equipment or tools. It's a broad church, basically.

Why Do Technical Skills Matter?

These skills matter because they're essential to getting the job done well. If an accountant lacks the skills and knowledge to properly manage a business's finances or file tax returns, they won't be a successful accountant, no matter how great they are at the "softer" skills like communication and collaboration.

(Let me stress that the inherently human "soft" skills such as creativity, communication, collaboration, and decision-making will still be highly prized in the workplace, and you can read more about these must-have skills in later chapters.)

The fact that technical skills matter likely isn't news to you. It's probably obvious that you need to be proficient in the practical skills related to any job. So you might be wondering why I've devoted a chapter of this book to the topic. It's because we're reaching a crucial tipping point in the nature of work—the point where technology is advancing faster than ever and more human tasks are being given over to machines. Far from making technical skills less important, I believe this rapid evolution will make technical skills more important than ever—albeit the specifics of those skills may change as technology evolves.

In fact, the most in-demand jobs in the world all require some form of technical skill. As of 2021, these most in-demand jobs included nurse, teacher, physical therapist, construction worker, web developer, and financial advisor.[1]

If we're brutally honest, those jobs that don't require much in the way of technical skills will ultimately disappear. These are the jobs comprised of easily repeatable, easily automated tasks—tasks that are usually better suited to machines. Let's take a checkout worker in a grocery store as an example. I mean no disrespect to checkout workers, and I know so many customers value the human interaction with a cashier. But if we look purely at the practical side of the job—scanning items and facilitating payment—we already know those tasks can be done by machines. Walk into any mid-sized or large grocery store, and you'll see self-service checkouts. In many stores in the UK, these self-service checkouts outnumber regular checkout workers, especially at off-peak times. And in the US, Amazon has introduced highly automated grocery stores, called Amazon Go, that have eliminated checkouts altogether.

In the long run, we may not have any human checkout workers at all. We need only look at obsolete jobs from the past, such as elevator operator, film projectionist, and VCR repairperson, to know that certain jobs die out over time as technology evolves. And in the future, anything that can be easily automated, will be. This means a whole host of jobs may ultimately disappear, including things like travel agents (thanks to online travel booking sites), telemarketers (hello, annoying automated marketing calls), taxi drivers (driverless taxis are already in use in Phoenix, Arizona), sports referees (thanks to video assistant referee [VAR] systems), and even bookkeepers (given the pace of change in bookkeeping software).

I cannot stress enough that the jobs that remain, and those new jobs that are created, will all undoubtedly be affected and augmented by technology in one way or another. A hairstylist in a salon, for example, is unlikely to be replaced by a robot hairstylist anytime soon, but that doesn't mean the job won't be augmented by new technologies. All professions are evolving constantly, and hairstyling is no different. In the not-too-distant future, your stylist will be able to use augmented reality mirrors—mirrors that overlay a digital image over your real-life reflection—to show you what your new hairstyle will look like before they've snipped a single lock of hair, or let you see whether a new color would suit you before you physically take the plunge. The technology exists and is already in use in certain salons.[2]

Likewise, truck drivers are learning to work with autonomous vehicle technology that can take on some or all of the driving responsibility, at least on highways if not in depots and busy urban areas. Radiologists are aided by AI-enabled scanners that can interpret scans and carry out routine reading and measurement tasks on medical imaging, freeing up radiologists to work on more complex diagnostic cases and to have more time for the treatment and management of diseases. Accountants are already leaving many of the basic financial tasks to software and instead focusing more on business advisory services that are geared towards helping business owners run better, more profitable businesses. Plumbers and electricians increasingly need to be able to work with smart meters and renewable technologies. Farmers can now use an array of tools that augment or automate farming practices—from deciding what to plant and when, to picking fruit. Data scientists can deploy AIs for the nuts-and-bolts analysis work, while focusing their time on more strategic data tasks.

Although almost all jobs will be augmented by new technologies, we'll still need people with technical knowledge and skills—from job-specific professional skills to the digital and data skills I talked about in Chapters 1 and 2. We don't want to lose crucial technical skills. If there were no human radiologists anymore, for example, who would further the field of radiology and ensure diagnostics and treatment continues to get better? Having people with technical skills, whatever those skills may be, is what allows any industry or sector to evolve. It's how we drive industries forward and build the solutions of the future. And it's how we get the best out of technology.

What Are the Essential Technical Skills?

There's no one-size-fits-all answer to this question, since essential skills will vary greatly from job to job. The key takeaway here is that technical skills will continue to be important—indeed, will grow in importance—in our rapidly changing world.

In terms of key skills, broadly speaking, we're talking about the non-IT professional skills required for your job, as well as the skills needed to work alongside technology. These include the digital literacy skills I talked about in Chapter 1, the data literacy skills from Chapter 2, and the cyber-threat awareness skills coming up in Chapter 4, plus the practical skills applicable to your job or the career you want to move into—whether that is bricklaying, medical image analysis, or being able to drive an 18-wheeler truck.

All of those skills aside, some of the essential technical skills for 21st-century work are likely to include:

- Customer relationship management

- Project management

- Social media management

- Video and other content creation

- Product development and product lifecycle management

- Technical writing, or being able to explain complex subjects in plain English

- Mechanical maintenance

Obviously not all of these will apply to all jobs, but it gives an idea of the (non-IT) technical and practical skills that many employers are likely to value.

Looking specifically at IT and technology jobs, some of the most in-demand technical skills will be around:

- Programming languages
- AI and machine learning
- Data science, data analysis, and data visualization
- Cybersecurity
- Cloud computing
- 5G
- Internet of things
- Software development
- User experience (UX)
- Extended reality (augmented reality, mixed reality, and virtual reality)
- Robotics
- Quantum computing
- Blockchain

Again, I'm not talking about technology professionals needing to be proficient in all of these skills. Each are specialisms in their own right. That said, a basic knowledge of all tech trends will be increasingly important as technologies converge and influence each other.

How to Improve Your Technical Skills

The evolution of professional and technology-related skills is faster than ever, which makes keeping up a bit of a challenge (and that's putting it mildly). We must all expect fast innovation and rapid change. But this also presents a great opportunity for everyone. In the 21st-century workplace, those who possess and maintain technical skills will no doubt be more desirable to employers. And for businesses, investing in technical skills is a vital part of driving business success and staying ahead of the competition.

As for how to develop technical skills, that will depend on the specific job and industry. But let's look at some broad tips for individuals and employers.

For individuals

You absolutely should encourage your employer to invest in the technical training needed to do your job. On top of that, as with all of the skills in this book, taking an active learning approach is essential for learning new skills and keeping up with changes in your field. Depending on your area of expertise, this may include:

- Signing up for online courses. Whatever your chosen field or desired career path, there will no doubt be useful courses online, through providers like Coursera and Udemy. For example, Udemy has courses on everything from electrical wiring to making YouTube marketing videos for businesses.

- Using self-study materials such as books and podcasts to learn about new topics or keep up with the latest technical trends in your industry.

- Learning from a professional in the same field. From mentoring to job shadowing, spending time with an expert is a great way to pick up practical and technical skills.

And always be sure to include your in-demand technical skills on your resume (something that goes for most of the skills in the book!).

For employers

It's vital employers invest in specialized learning programs and on-the-job training to equip their people with the technical skills needed to drive the business forward. There are many ways of going about this; you just need to find the ways that work best for your business, your people, and the required skills. But here are some general tips for getting the best out of technical training:

- Set your training and development goals. Here, it's really important to talk to managers and team members to discover where there are learning gaps. Only then can you build a development program that truly meets the needs of the business and its people.

- Emphasize the benefits that technical training will bring to individuals (for example, how it will help with their current role, future prospects, and so on).

- Make use of online learning and self-study materials that let people learn at their own pace. For some people, bite-sized learning may be a better way of absorbing technical information.

- Gamify learning where possible. Make learning more fun and motivating by using strategies such as points, levels, and leaderboards.

- Consider whether augmented reality and virtual reality can help to bring your technical training to life. BP, for example, has used VR to train oil refinery workers on emergency procedures. (If this area is of interest, you might like to check out my book *Extended Reality in Practice,* which features lots of examples of VR- and AR-enhanced training.)

- Build a culture of curiosity and active learning, where learning new things is seen as an opportunity rather than a burden.

Key Takeaways

In this chapter, we've learned that:

- Despite the rapid changes in our working lives, and increasing digitization, technical skills will still be an important future skill.

- In general, this means investing in the practical and physical skills needed to do a job effectively, as well as the skills needed to work alongside technology.

- Remember, this is an area where skills will grow stale relatively quickly, especially as new technologies come on the scene—and this will apply to almost all industries (consider our hairstylist and farmer examples from earlier in the chapter). So you must be prepared to continually refresh your technical skills.

Before we move onto the essential softer skills, there's one more technical topic that we need to address: cyber-threat awareness. Let's explore how every employee, at every level of the business, must take responsibility for keeping individuals and companies safe.

Notes

1. Top 25 In-Demand Careers in 2021; Indeed; https://www.indeed.com/career-advice/finding-a-job/in-demand-careers

2. I went to Amazon's high-tech hair salon and virtually dyed my hair pink—then got the best haircut I've ever had; Business Insider; https://www.businessinsider.com/amazon-salon-hair-dye-pink-virtual-app-london-2021-9?r=US&IR=T

CHAPTER 4
DIGITAL THREAT AWARENESS

As the world becomes more digital, so too do the approaches taken by criminals and other nefarious characters. And this brings with it many digital threats that impact our personal and professional lives—from having our personal account credentials stolen to organizational systems being hacked. On top of this, there are issues like digital addiction on the rise.

In this chapter we'll look at the biggest digital threats that apply to everyday life, then move onto the key techniques employed by cybercriminals. Because we're covering multiple digital threats within the same chapter, you'll notice I've included the "how-to" tips in with each section, rather than in a standalone section at the end of the chapter.

What Is Digital Threat Awareness, and Why Does It Matter?

Digital threat awareness essentially means being aware of the dangers of being online or using digital devices—and having the tools to keep yourself (and your organization) safe. This is vital because more and more of our lives involve a digital element. We bank online. We shop online. We use digital communication tools. We read the news online. We track our personal fitness with digital tools. We order the Friday night takeaway from our phones. We receive updates from our children's schools via social media. We make doctors' appointments and order repeat prescriptions online. Our digital footprints are bigger than ever.

And this digital footprint spans both personal and working lives, not to mention the widening grey area in between the two. (Think of taking your personal devices into work and connecting to the company network, or accessing company systems from home.) With more and more of us working from home, this blurring of our work and personal digital lives is likely to continue. And this means we need to be constantly vigilant about digital threats.

The Biggest Digital Threats in Everyday Life

It's obvious that so much of everyday life now involves a digital element. And this brings with it certain dangers that we all need to be aware of. If you're a parent, you'll need to be especially on your guard, because not only do you have to educate your children about digital dangers, but you'll also have to watch for warning signs that something might be going wrong.

Here are the biggest digital threats that I believe will impact life in the 21st century.

Digital addiction

This can span any type of addiction related to digital pastimes, including social media addiction, internet addiction, phone addiction, and gaming addiction.

The stats on digital addiction are bleak, particularly when it comes to social media. Worldwide, an estimated 210 million people suffer from social media addiction.[1] And it's no wonder: social media sites like Facebook and Instagram are literally designed to be addictive. Ex-employees of big tech giants have admitted as much. After all, the more time we spend scrolling and tapping, the more companies like Facebook earn in advertising revenue. And so these apps are designed to maximize the amount of time we spend on them—with astonishing effectiveness. American teens spend an average of nine hours a day on social media; and adults aren't much better, with 50 percent admitting they've used social media while driving.[2]

But it's not just social media sites that are addictive. Smartphones have us constantly coming back for more. Americans now check their phones on average 262 times a day (that's every 5.5 minutes), with 80 percent reaching for their phone within 10 minutes of waking up. So I wasn't surprised to read that 75.4 percent of Americans consider themselves addicted to their smartphone, and 43 percent say their phone is their most valuable possession.[3] (More surprising perhaps is the fact that 45 percent of people would rather give up sex for a year than give up their smartphone.) True, much of the time we spend on our phones is spent scratching that social media itch, but there's no doubt that the phone itself has become something we turn to for validation and distraction—so much so that Dr. Anna Lembke, a world leader in the treatment of addiction, refers to the smartphone as "the modern-day hypodermic needle."[4] As she puts it, we now struggle to be alone with our thoughts, and instead prefer the instant dopamine release that comes from reaching for our phones. "We're losing our capacity to delay gratification, solve problems and deal with frustration and pain in its many forms," she says.

So how can you spot the signs of digital addiction? Depression is a key indicator, especially for social media addiction. Teens, for example, are twice as likely to exhibit symptoms of depression when they spend more than five hours a day on their smartphones.[5]

Thankfully, there are many practical strategies we can all employ to limit the amount of time we spend on our phones. You can set time limits for apps in your phone's settings. You can set a "bedtime" mode, where your phone turns off notifications between the evening and the morning—or simply turn off app notifications altogether. If you want to be really strict, you can delete the most distracting apps from your phone (you can always log in on your computer if you feel a desperate urge to see Cousin Gary's holiday pictures).

Then there's helping your children to build healthy (or health*ier*) digital habits. For example, in my family we have a "one-screen" rule, which means literally one screen at a time—so no going on their phones during family movie night. (My wife and I follow the same rules!) We also try to teach our children that it's okay not to log in to an app every day. Gaming apps can be particularly pernicious, encouraging players to log in daily to complete their "login streak" to gather extra points or the like. This caused my 10-year-old son to feel pressured to log in just for the sake of logging in. We had to explain to him that it's okay to break that streak.

Online privacy, digital devices, and the need to protect your data

Your data is a precious commodity, and the chances are you're earning money for companies without even knowing it, simply by using their app or service.

As a rule of thumb, the free apps are the worst for gathering and selling masses of data. In other words, if it's free, you can be pretty sure you (or, more accurately, your data) is the real product. You'd be surprised how many apps and services are harvesting your data and selling it on to third parties. Social media platforms are among the worst culprits. No surprise there. But did you know that food delivery apps like DoorDash, Deliveroo, Caviar, and Uber Eats rank right up there with social media sites when it comes to gathering and selling personal data?[6, 7]

What sort of data are we talking about? Not just your name and location. Depending on the app, it could be your email address, health and fitness data, financial information, browser history, purchase history, and even your contacts. Of course, sometimes this info is purely used by the company gathering the data, but in many cases—in as many as 52 percent of apps[8]—that data can be sold on to other companies who then use it to build up a clear profile of you and the products you're most likely to purchase. (As an example, Apple has come under fire for its most popular apps gathering user data while people sleep and sending that info to third parties—info including location, emails, phone numbers, and IP addresses.[9])

This data can be turned directly into corporate profits. Take Cosmose AI for example. Cosmose buys data from more than 400,000 apps and 1 billion smartphones, and shares detailed insights about consumer behavior with some of the world's biggest brands, including Walmart, which in turn allows those brands to sell you more products.[10] The good thing about Cosmose is the data is anonymized, but it still shows how valuable our data is (Cosmose is valued at $100 million), and how little most people understand about the sheer volume of data being gathered.

Then there's the plethora of smart devices now plugged into our homes, such as smart doorbells and smart speakers. These, too, are gathering data and potentially sharing it with others. This is incredibly personal data—think of all the things you say and do in your home.

While it's true that our personal data is generally used to make valuable improvements to the products and services we use (Alexa voice recordings, for example, help Amazon troubleshoot commands and increase Alexa's vocabulary), we also need to remember that our private data is a valuable product—and stop giving it away for free without thinking. Personally, before signing up for any new app or buying a new smart device, I always comb through the terms and conditions to check I'm happy with the privacy settings. The two biggest questions to answer here are what data is gathered and who is that data shared with? I do the same for the many apps my children want to use. (Apple makes this easy for me by applying privacy labels to apps available on Apple devices, and these labels tell me at a glance what info the app will collect.) The same goes for browsers—in fact, I'd recommend browsing in "private" or "incognito" mode whenever possible, because this means third parties won't be able to trace your browsing history (although your internet service provider—and employer, if browsing on a company device—can still see your activity).

In many cases, you can also ask your smart home devices either not to listen constantly for activation phrases, or delete recordings they've made. (You can ask Alexa, for example, to delete everything you've said in the last day.)

Passwords

You'd be amazed how many people have something like "12345" or the word "password" as their password, and how many use the same password for everything, despite the clear warnings not to. If that sounds like you, take this as a wakeup call. It's time to start taking password security seriously.

Consider this frightening stat: thanks to the thousands of data breaches that have occurred in recent years, more than 15 billion stolen account credentials—meaning usernames and passwords—are available for sale on the dark web, with such account information granting criminals access to everything from social media and financial accounts to admin accounts for organizational IT systems.[11] What's more, using these lists of stolen credentials, hackers can easily try using the same info to access your other accounts (known as "credential stuffing")—which is why it's so important not to reuse the same combination of user ID and password. Hackers can also use phishing techniques to trick people into giving up their login information (more on phishing coming up later in the chapter). There's also a technique known as "brute force," in which AIs work through potentially billions of combinations to work out the correct password, and the "password spraying" approach, where the system simply tries commonly used passwords like 12345 against user account names.

AI means this can all be done automatically now. In other words, we're no longer talking about the stereotypical "loner in a dark room" trying to guess your password—machines are cracking passwords effortlessly using a range of techniques.

So what to do about it? The most basic defense is to create robust passwords that have at least eight characters (12 is even better), using a mixture of cases, special characters, and numbers. And to do this for every account (rather than reusing passwords). Avoid predictable passwords that can be easily guessed from a glance at your social media profile. (While we're on the subject, avoid sharing too much personal information on Facebook and other social media sites, and ensure your profiles are visible only to friends.)

Even better, you can use a random password generator to create passwords that are harder to guess in a brute force attack (Chrome, for example, can suggest strong passwords for you). I'd also recommend using a password manager tool, such as Google's Password Manager, to securely store all your unique passwords. In addition, you can use two-factor authentication to secure your accounts—so even if a password is compromised, criminals still can't access your accounts.

Cyberbullying

Sadly, I think cyberbullying is now the most prevalent form of bullying that takes place. According to leading UK bullying charity, Bullying UK, 56 percent of young people have seen others be bullied online, and 42 percent have felt unsafe online.[12] With three children of my own, I am hugely concerned about this—especially since cyberbullying can happen 24 hours a day, seven days a week, and can even go viral. That's why we all need to be able to recognize cyberbullying, to teach our children what cyberbullying looks like, and to take appropriate action when it occurs.

But what counts as cyberbullying? It's basically any form of bullying that takes place online or through smartphones or tablets, which includes mediums like text messages, Snapchat, WhatsApp, and other messaging services, gaming sites, social media platforms, chat rooms, and message boards. Cyberbullying can take the form of harassment (such as sending offensive messages, or making nasty comments on social media posts), sharing damaging photos or information about another person, making threats, spreading fake rumors and gossip, stalking, intentionally excluding someone from online activities, and even impersonating someone online.

Parents should be on the lookout for signs that their child is being cyberbullied—signs such as becoming noticeably upset after being on their phone, showing signs of depression, not participating in activities they previously enjoyed, not engaging with family and friends, and a drop in school grades.

So what can we do about cyberbullying, aside from never going online (which isn't an option for most of us or our children)? The first step is to know your rights, in particular whether cyberbullying is a criminal offence in your country or state. In the UK, for example, it's against the law to use the phone system—which includes the internet—to cause alarm or distress. Specific activities may even fall under the UK's 1997 Harassment Act. Threats of harm should always be reported to the police, using screenshots as evidence. (As a general rule, always keep a log of evidence of cyberbullying because you never know when you may need it.)

Where children are involved, you should absolutely report cyberbullying to the school. In many jurisdictions, schools are legally required to take action; all US states, for example, have laws that require schools to respond to bullying, and many states now include cyberbullying under these laws.

In the case of social media and messaging boards, you (or your children) can take action to block specific users outright, and/or report them to the platform in question. Or, for a more discreet approach, Facebook and Instagram both have a feature called Restrict that lets you block a specific user without that person ever knowing. (Basically, the bully will still be able to comment and see their comments on your posts, but you and other users will never see what they say.) Both sites also have settings that you can turn on to automatically filter offensive comments and DMs.

To get offensive or inappropriate content removed, you can either report it to the platform in question, or get help from bullying organizations to get things taken down. (There's an organization online called Remove Harmful Content, which is a good start.)

Finally, keeping your social media profiles private and secure (by which I mean with secure passwords) and not oversharing personal information and images are other good ways to defend against the various forms of cyberbullying.

Digital impersonation

As more of our lives go online (including images, videos, and recordings of us), digital identity theft is becoming more of a threat. Social media impersonation is a particular threat to watch out for. Here, fraudsters create social media accounts that use someone else's (or even an organization's) name, image, and other identifying features to create fake accounts. Indeed, I've experienced this myself, with my public photos being used to create fake (but authentic-looking) Facebook accounts in my name. Even if your identity isn't used by fraudsters, there's still a risk that you could be interacting with fake accounts online.

Why would anyone want to create an account in someone else's name? In some cases it can be related to a vendetta or stalking. Or it can be part of a wider scam to filch other users' money or personal data (particularly when it's a brand or a figure in the public eye that's being impersonated).

Social media impersonation is also a key feature of catfishing—where fake identities are used to form dishonest relationships online. Catfishing can be used for revenge and harassment, or it may simply be because the person doing the catfishing lacks confidence in their real identity, or has mental health issues. Whatever the underlying reason, the effects of catfishing can be devasting to the victim, especially if they were emotionally invested in the relationship.

Spotting fake accounts online can be tricky, but it's something we must all become accustomed to in our increasingly digital world. As a general rule, fake accounts may be recently created, with few friends or images on their profile. To avoid being targeted by fake accounts on social media, you can adjust your privacy settings so your profile isn't public, and only friends see your posts. And whenever you do accept new friend or follower requests, or follow a new account yourself, be vigilant; don't be rushed into sharing personal information or images, and never give money to anyone who asks for it online.

If you're concerned that your identity might be used by others, do a regular search of your name and look for images of you. (A reverse image search lets you upload your images and find out where they exist online.) Also think very carefully about the information you share on social media—personal information, photos, and so on—because it can all be used to create authentic-looking accounts in your name. This also ties back into the password security, since thieves can use information gleaned from social media accounts either to steal your login credentials or to attempt to change your password without your knowing. So think twice before taking those ubiquitous online quizzes and sharing things like your mother's maiden name, pets' names, your high school mascot's name, or where you met your partner—all information that is commonly used as security questions to change passwords on accounts.

Understanding Key Cyber Threats

Now let's move onto specific techniques that cybercriminals commonly use to infiltrate individual or organizational systems, or harvest sensitive information such as passwords.

Data breaches

Data breaches, or the theft of data by a malicious actor, are a huge problem for organizations—not to mention the individuals who have trusted those organizations with their personal data. The billions of login credentials available for sale on the dark web show the sheer extent of data breaches, which are often the result of a phishing scam or malware, and several other concerns in this realm.

Phishing

In phishing, scammers target victims with spam messages (usually emails, but sometimes through SMS and other messages) that prompt the reader to take urgent action, such as changing a password. The email may contain a link that takes the recipient to a fake (but potentially legitimate-looking) version of a website, often with the goal of nabbing their username and password or financial details. Alternatively, the email may contain a malicious attachment or link that infects the target's system with malware. Unfortunately, phishing is now so common and sophisticated that it can be difficult to spot a fake message when it lands in your inbox, so you need to be extremely wary of clicking on any links or attachments sent to you (more on this coming up later).

Malware

There are many types of malware—malicious software—with the Trojan being perhaps the best known. Malware is essentially a piece of malicious code that is planted on the target's computer or network to perform a specific activity, such as gathering sensitive information from the system, or infecting the system with a virus. Typically malware enters the target system when a user clicks on something that may look entirely genuine, such as a software update, or an innocuous link or advertisement.

Ransomware

A particularly damaging form of malware, ransomware is where files on the target's system are encrypted, and can no longer be accessed. The target basically has to pay the perpetrator a ransom to regain access to their own systems. This can be very lucrative for criminals; according to cybersecurity firm Coveware, the average ransomware payout grew to almost $234,000 per event in 2020.[13] Increasingly, ransomware attackers are using a technique called *data exfiltration*, where the target's data is copied as well as being encrypted—and the attackers threaten to release the data publicly if the ransom isn't paid (and sometimes even if the ransom is paid).

IOT attacks

Thanks to the huge rise in smart Internet of Things devices in our homes and offices—plus the fact that these devices are often unsecured and operating on out-of-date software—hackers can now use these devices to cause havoc. In particular, hackers can harness thousands of these devices at once and use them to overwhelm target systems, such as websites, with fake traffic (known as a distributed denial of service, or DDoS, attack), causing the system or site to crash from overload. This is usually done to cause disruption or embarrassment, rather than to steal data.

Defending Against Cyber Threats

Here are some tips for individuals:

- Be suspicious of emails, especially if they come from unknown sources or odd-looking email addresses (check the sender's email address, not just the display name, which is easily forged). Be especially wary if the email is pressuring you to do something, such as reset an account or change your password. Phishing emails prey on our panic responses, and rely on us quickly clicking on a link and giving over our sensitive information. If you aren't sure whether a request is genuine, contact the organization (for example, your bank) or individual directly via the usual channels (not using any contact details or links in the email).

- Never open attachments or click on links in emails from sources that you don't 100 percent trust. Personally, I never click on links in emails anyway, even if it's a genuine email from a company I trust. I'll log in to my account in my usual way, via my browser or the app in question.

- Never click on pop-ups, which are very often linked to malware. Even better, install a pop-up and ad blocker.

- Don't input sensitive information or download files from websites that don't have a security certificate. On secure sites, the website URL will start with "https" or have a closed padlock icon next to the URL.

- Ensure all devices run the most up-to-date versions of software. This includes your smart home devices (which should also be password protected with a unique password, not the default system password).

- Practice good password hygiene by following the tips earlier in the chapter.

- Only download apps from trusted sources.

- Install a firewall and antivirus software.

- Always back up your data regularly.

- Try to avoid using public Wi-Fi networks where possible. If you do use them, use a secure VPN service to keep your connection private.

Tips for organizations

The same tips that apply to individuals generally apply to organizations as well. But to successfully implement these tips at an organizational level, you'll need a company culture that prioritizes cybersecurity, which means you need to train your people so they can identify threats and know what to do when they spot something suspicious.

In addition to this, you should:

- Use threat-detection tools as part of your cybersecurity toolkit.

- Regularly test your systems to ensure they are up to date and protected.

- Have a plan in place that sets out what happens in the event of a security breach. This should cover the technology response (how you plan to stop the attack and secure systems), the people response (who you need to tell), and the post-breach investigation (learning from the attack so it doesn't happen again).

This is by no means an exhaustive list—more a reminder to take steps to protect your organization and raise awareness of cyber threats among employees.

Key Takeaways

In this chapter, we've learned:

- In everyday life, there are several digital threats that we all need to be aware of, with the biggest being digital addiction, data privacy, password theft, cyberbullying, and digital impersonation.

- On top of these everyday dangers, there are specific cyber threats that individuals and organizations must defend against, including data breaches, phishing, malware, ransomware, and attacks on IOT devices.

- Thankfully, there are several steps individuals and organizations can take to protect their systems, such as practicing good password hygiene and harnessing antivirus and firewall tools (to name just a couple).

- Over and above this, we all need to learn to develop a healthy—or health*ier*—relationship with technology, and teach the next generation how to get the best out of technology without being dominated by it.

For me, staying safe in the digital world also means flexing my critical thinking muscles. Challenging the things we see online, and asking questions such as "Is this genuine?" and "Do I trust this source?" will help us all avoid some of downsides of technology. So let's head over to the next chapter and explore essential critical thinking skills in more detail.

Notes

1. 17 Social Media Addiction Statistics; Truelist; https://truelist.co/blog/social-media-addiction-statistics/

2. Ibid.

3. Cell Phone Behavior in 2021: How Obsessed Are We?; Reviews.org; https://www.reviews.org/mobile/cell-phone-addiction/

4. Constant craving: how digital media turned us all into dopamine addicts; *The Guardian*; https://www.theguardian.com/global/2021/aug/22/how-digital-media-turned-us-all-into-dopamine-addicts-and-what-we-can-do-to-break-the-cycle

5. 17 Social Media Addiction Statistics; Truelist; https://truelist.co/blog/social-media-addiction-statistics/

6. These popular apps collect the most data about you; TechAdvisor; https://www.techadvisor.com/news/security/apps-collect-personal-data-3805030/

7. Social Media and Food Delivery Apps Sell the Most Personal Data; PC Mag; https://www.pcmag.com/how-to/social-media-and-food-delivery-apps-sell-the-most-personal-data

8. Ibid.

9. Hidden iPhone trackers harvest data while you sleep; Komando; https://www.komando.com/security-privacy/hidden-iphone-trackers-harvest-data-while-you-sleep/569770/

10. Smartphone Tracking Data And Artificial Intelligence Turn People's Movements Into Detailed Insights and Profits; *Forbes*; https://www.forbes.com/sites/bernardmarr/2020/10/07/smartphone-tracking-data-and-artificial-intelligence-turn-peoples-movements-into-detailed-insights-and-profits/?sh=2ae000716bef

11. Billions of stolen passwords for sale on the dark web; WeLiveSecurity; https://www.welivesecurity.com/2020/07/09/billions-stolen-passwords-sale-dark-web/

12. Cyber bullying; Bullying UK; https://www.bullying.co.uk/cyberbullying/

13. Ransomware demands continue to rise as data exfiltration becomes common; Coveware; https://www.coveware.com/blog/q3-2020-ransomware-marketplace-report

CHAPTER 5
CRITICAL THINKING

Now we move from hard, technical skills onto the soft skills that are related to personality and personal attributes. But just because something is a soft skill doesn't mean it can't be learned and improved. Any soft skill can be honed. And hone we must, because as more and more practical tasks become automated and given over to machines, certain soft skills will become increasingly valuable in the workplace.

For me, critical thinking is right up there as one of the most vital soft skills to cultivate for future success. In this era of fake news and social media bubbles—and indeed, with the sheer volume of information that we're now bombarded with on a daily basis—the ability to look at evidence, evaluate the trustworthiness of a source, and think clearly is becoming more important than ever. Critical thinking is often confused with being critical or negative, but that's a misconception. Critical thinking is about *objectivity*. It's about having an open, inquisitive mind—something that all employers find valuable.

Critical thinking isn't just important for career success (or academic success, for that matter). If you lack the ability to think critically, you may be more vulnerable to things like manipulation, fraud, and fake news in your everyday life. Critical thinking is an important life skill, then—and not just something you get credits for at university or a throwaway statement that looks good on your CV.

What Exactly Is Critical Thinking?

We all think, right? As humans, we can't help but think. (If you've ever tried to meditate, you'll know how hard it is to "turn off" your thoughts.) But there's a big difference between the kind of thinking we do routinely and thinking *critically*. Because our everyday thinking—the kind of thinking we do without, well, thinking about it—is hampered by things like incomplete information, personal opinions, assumptions, biases, and even prejudices. Not all thought is high-quality thought, in other words. This is why we need critical thinking.

Critical thinking essentially means thinking objectively. It means analyzing issues or situations based on evidence (rather than personal opinions, biases, and so on) so that we can build a thorough understanding of what's really going on. And from this place of thorough understanding, we can make better decisions and solve problems. To put it another way, critical thinking is Mister Spock coolly applying reason and logic to an unfolding situation, while Bones is in the background spluttering out the first emotion-laden thought that comes into his head.

Delving a little deeper, critical thinking is comprised of several processes:

- Observing and identifying the problem or situation.

- Gathering the relevant facts, and analyzing the sources of information. A critical thinker will ask themselves whether they're seeing the complete picture, and what information may be missing.

- Recognizing biases that may influence the claims and arguments of others (but also recognizing your own biases).

- Spotting inconsistencies and errors in arguments.

- Being curious enough to ask questions, rather than take everything at face value. At some point we lose that childhood trait of constantly asking "But *why?*" Part of critical thinking is retraining ourselves to ask such open-ended questions.

- Figuring out which arguments and information are relevant and important, and which are not.

- Drawing (or inferring) conclusions based on the information available.

- Considering the consequences of various choices, solutions, and outcomes.

- Solving the problem or conflict, or deciding on next steps, based on all of the previous processes.

In a nutshell, I like to think of critical thinking as *active, independent thinking*—as opposed to passively swallowing information and taking it at face value. Importantly, this is a skill that can be cultivated and improved over time, meaning we can all take steps to become better critical thinkers. And, given some of the challenges in the modern workplace and the world at large, it's vital we all do just that.

The Problem Today, or Why We All Need Critical Thinking Skills

There are some troubling trends at play in society today—trends that pose a threat to critical thinking and, at the same time, make critical thinking more important than ever. At the end of the chapter, we'll look at ways to overcome these roadblocks and improve critical thinking, but for now let's focus on the trends themselves.

Cognitive biases

Bias obviously isn't a trend. Humans have always had biases (and, yes, we all have them). But trends such as increasing polarization and social media filter bubbles can make it harder to recognize and break free of biases. That's why it's worth dwelling on the subject of bias, and how it affects thought.

No matter how rational and logical we think we are, the truth is we're all under the influence of cognitive biases—biases that influence our beliefs, thoughts, and decisions. Indeed, a good many of us will have experienced cognitive bias in action. Sometimes these can be very obvious, such as gender bias and stereotyping, but other types of bias can be incredibly subtle and hard to spot. For example:

- In cultural bias, we may perceive other cultures as being somehow less "normal" than our own.

- Confirmation bias is the tendency to focus on information that supports something we already believe—to the extent that we dismiss or avoid information that may disprove what we already believe.

- Anchoring bias is the tendency to be overly influenced by the first piece of information we hear.

- In hindsight bias, we perceive events—even random events—to be more predictable *after* they have happened (often referred to as the "knew it all along" phenomenon). As a result, we may overestimate our ability to predict events in future, and take unwise risks.

- In selection bias, we notice things more because something has happened to *make* us notice it more. A good example is buying a new car and then suddenly noticing that same make and model everywhere on the roads.

- In the halo effect, an initial impression of someone influences what we think of them overall (for example, thinking an attractive person is smarter or more capable than average). This can have a huge influence in areas like hiring.

- The flip side of the halo effect is the horn effect, where one trait that you perceive as negative can influence your entire opinion of someone.

These are just a few of the biases that can have a powerful impact on our everyday thought and decision-making. There are many more. On top of these cognitive biases, there's often bias in the data we use to make decisions (see Chapter 2), meaning the data may be skewed towards a particular group, or exclude or underrepresent others. All of which means we need to work hard to spot bias in our own thinking, and in the information we're presented with.

Increasing polarization

The world seems to grow smaller every day as the gaps between our different cultures lessen (think of British teens soaking up South Korean pop music, or Americans celebrating Saint Patrick's Day). And yet, at the same time, we seem to be getting more, not less, divided. This is the polarization trend, where people are divided into sharply opposing camps, with little or no ability to identify with the other side. Think of the Brexit versus Remain debate, or the apparent chasm between Democrat and Republican voters in the US.

One 2018 study perfectly demonstrates polarization in action. In it, people were asked to estimate the percentage of Democrats who are Black, atheist or agnostic, and gay, lesbian or bisexual, and the percentage of Republicans who are evangelical, 65 or older, Southern, and earning more than $250K a year—essentially playing on stereotypes for both parties. Democrats believed that 44 percent of Republicans earned more than $250K a year (the true figure is 2 percent), and Republicans believed that 38 percent of Democrats were gay, lesbian, or bisexual (when it's really around 6 percent).[1] In other words, misconceptions ran high when describing the "other" party. And the more political information people consumed, the more they were mistaken about the other side.

How can this happen? Fundamentally, the way we consume information has changed. The internet has made it possible to access whatever information we want and yet the sheer vastness and choice of information has only widened gaps in society, particularly when it comes to politics (although political polarization isn't the only kind of polarization). As Ezra Klein puts it in his book *Why We're Polarized*, "Greater choices let the devotees learn more and the uninterested learn less."

Based on this, you might think that the cure for polarization is consuming a wider variety of information, including info from the "other side," but that doesn't seem to help. One study had Twitter users who identified as either Democrat or Republican follow a bot that shared tweets from authoritative figures and organizations on the other side. The participants were regularly surveyed about their views during the month-long experiment, and these surveys showed that respondents became *more* polarized, not less, after being exposed to opposing voices.[2] Republicans expressed substantially more conservative views after the month was up, and Democrats became slightly more liberal.

So what can we do about this? The first step is to recognize that we're all prone to polarization, and not just political polarization. If you've ever clicked on a Buzzfeed headline along the lines of *22 Signs You're a '90s Child* or *33 Things Only People with IBS Will Understand*, you'll know how alluring it is to feel part of a "tribe." This is why we need to exercise critical thinking skills in all walks of life. By thinking critically, we can spot the information that attempts to position different groups as "other," learn to question our assumptions, and apply logic to the information we consume on a daily basis.

Social media filter bubbles

One reason for the increasing polarization in our world is the fact that so many of us get our news from social media—especially younger people. More than half of teens, for example, say they get their news from Instagram, Facebook, and Twitter.[3]

The trouble with social media is that it's designed to keep us coming back for more, and as such these platforms repeatedly serve up information that they know we'll like, based on our existing interests and beliefs. (By the way, I highly recommend you watch *The Social Dilemma* on Netflix. It shows with astonishing clarity how social media apps are designed to maximize our attention.) So if you like and share anti-vaccination content on Facebook, for example, the platform will show you more and more of the same content. The danger is we can end up thinking the world is exactly as we see it online because we're never presented with information that challenges our beliefs.

This is what's commonly known as a "filter bubble"—a state of information isolation caused by algorithms feeding us content that we agree with, based on our previous behavior. In the filter bubble, news that we dislike or disagree with is automatically filtered out, and this can create an "echo chamber" effect, where our perception of reality can become distorted because we're only ever exposed to views that mirror our own.

Critical thinking helps by encouraging us to step beyond the filter bubble and echo chamber. Someone who thinks critically is able to stop and ask themselves "Am I seeing the world as it really is, or am only seeing a narrow slice that echoes my own beliefs?"

Fake news

There's also the fact that what we're seeing on social media may not even be true! Misinformation spreads like wildfire online, and two-thirds of US adults say they've come across false information on social media. If, like me, you're surprised that figure isn't higher, consider the sobering statistic that 56 percent of Facebook users can't actually recognize fake news that aligns with their beliefs.[4]

Misinformation was particularly rife during the pandemic, prompting the World Health Organization to use the term "infodemic" to describe the rapid spread of misleading or fabricated news.[5] As an example, according to a 2020 NPR/Ipsos poll, 40 percent of Americans believe COVID-19 was created in a lab in China, even though there is no evidence for this.[6]

While some of the misinformation online isn't necessarily a deliberate attempt to deceive people, we can't ignore the impact of malicious disinformation, designed to sow confusion and chaos. (Note the distinction between *misinformation* and *disinformation*. Both present untrue, inaccurate, or misleading information, but the difference lies in the intent. Misinformation refers to false or out-of-context information that's presented as fact, regardless of an intent to deceive people, while disinformation is a type of misinformation that intentionally and maliciously attempts to deceive people. To put it in regular terms, misinformation might be your Aunt Jan sharing an article on Facebook that she believed to be true, but you later find out to be false; while disinformation might be a political party or nation-state deliberately setting out to deceive or mislead people with false, incomplete or out-of-date information.)

To be clear, I'm not just talking about countries like Russia and China intentionally spreading disinformation; during the pandemic, President Trump repeatedly touted unproven medical treatments, contradicted his own government scientists, and even retweeted conspiracy theories by celebrities Diamond and Silk, whose own Twitter account was locked for spreading false information about the pandemic.[7] Trump ended up being permanently banned from Twitter, but undeterred, he was busy announcing his own social media app TRUTH Social just as I was writing this chapter. What really troubles me is that Trump's plans include a video-on-demand service that will feature entertainment, news, and podcasts.[8] This planned progression from social media to content creation is likely to make the whole polarization and echo chamber problem even worse, and make it harder for people to understand that other views exist—yet another reason why we need critical thinking skills.

The rise of deepfakes

Deepfakes are where artificial intelligence and deep learning techniques are used to create fake images, video, or audio—often with disturbingly realistic results. If you're not sure you've seen a deepfake, watch the YouTube video "Spot on Al Pacino impression by Bill Hader" and see the comedian's face eerily morph into a young Al Pacino as he mimics the actor's voice. Or check out @deeptomcruise on TikTok, an account devoted entirely to Tom Cruise deepfakes, in which you can watch fake Tom Cruise performing magic tricks or doing banal stuff like washing his hands.

A lot of deepfakes are pornographic (for example, mapping female celebrity faces onto porn stars' bodies), but they can also be used to embarrass public figures and even disrupt elections. Facebook took steps to ban deepfake videos that were aimed at misleading viewers in the run-up to the 2020 US election (basically, the videos were designed to make people think politicians had said words they did not actually say).[9] Other deepfakes are aimed at scamming people out of money. In one example, the boss of a UK subsidiary of a German energy company authorized the payment of £200,000 into a Hungarian bank account after falling victim to a fake phone call from the German head of the company, featuring what the company's insurers suspect was deepfake audio.[10]

Deepfake technology isn't inherently bad. Remember that digital avatar of me that I mentioned in Chapter 1? It's made possible thanks to deepfake technology. And in the future, as we spend more and more time in the metaverse (Chapter 1), the ability to have a super-realistic digital version of ourselves will mean you can hang out with friends online in a more immersive way, and even conduct meetings from the comfort of your sofa, with your digital self dressed in smart clothes, while the real you lounges comfortably in pajamas! But the flip side is that people could hijack our digital selves very convincingly, potentially with devastating results.

For me, the even bigger concern is that deepfakes will further undermine trust in what we see, which could of course be used to certain people's advantage. President Trump, for example, has reportedly suggested that the infamous audio of him boasting about grabbing women by the genitals was not real, despite admitting in 2016 that "I said it, I was wrong and I apologize."[11] In other words, as deepfakes become more of a problem, and become even more convincing, there's a distinct danger that political leaders will be able to claim things that really happened—things that we actually saw or heard—never really happened!

Bottom line, it'll become increasingly difficult to tell the real from the not-real, which means we all need to think more critically about the content we consume.

How to Improve Your Critical Thinking

It should be clear by now that individuals and organizations without critical thinking skills are at a disadvantage in today's world, so let's briefly explore some practical ways to beef up critical thinking skills.

For individuals

In essence, critical thinking means not taking information at face value. In practice, this means you should:

- Always assess new information. Whether it's an article someone has shared online or data related to your job, always vet the information you're presented with. Good questions to ask here include "Is this information complete?" "What evidence is being presented to support the argument?" and "Whose voice is missing here?"

- Consider where information has come from. Is the source trustworthy? What is their motivation for presenting this information? Are they trying to sell you something or provoke you to take a certain action?

- Gather additional information where needed. Where you identify gaps in the information or data, do your own research to fill those gaps.

- Ask open-ended questions. Critical thinkers are curious people, so channel that inner child and ask lots of "who," "what," and "why" questions.

- Find reputable sources of information. Personally, I turn to established news sites, nonprofit organizations (fullfact.org and factcheck.org are good places to start), plus education institutes—and try to avoid anonymous sources or sources with an axe to grind or a product to sell. (Remember, when you Google something, those very top results are ads, where the organization or website has paid to show up first. That doesn't make them an authoritative voice.) Also, be sure to check when the information was published.

- Try not to get your news from social media. And if you do see something on social media, check the accuracy of the story (via reputable sources of information, as above) before you share it.

- Learn to spot fake news. It's not always easy to spot false or misleading content, but a good rule of thumb is to look at the language, emotion, and tone of the piece. Is it using emotionally charged language, for instance? Also, look at the sources of facts, figures, images, and quotes. A legit news story will clearly state its sources.

- Learn to recognize and question bias. Biased information may seek to appeal more to your emotions than to logic, and/or present a limited view of the topic. So ask yourself, "Is there more to this topic than what's being presented here?" And don't forget about your own biases. Think objectively about your likes and dislikes, preferences and beliefs, and how these might affect your thinking.

- Form your own opinions. Remember, critical thinking is about thinking independently. So once you've assessed all the information, form your own conclusions about it.

Working with a mentor can help you practice these skills in your working life—at the very least, it can be helpful to have someone pull you up short when you're responding to information from a place of emotion or assumption. Also, do check out online learning platforms such as Udemy and Coursera for helpful courses on general critical thinking skills, as well as courses on specific subjects like cognitive biases.

For organizations

I'd highly recommend building critical skills training into your soft skills training programs—and ensuring critical thinking is covered in relevant technical training (especially data literacy).

Key Takeaways

In this chapter, we've learned:

- Critical thinking refers to the ability to think objectively and independently. It means being curious about the information we're bombarded with on a daily basis, and not taking everything at face value.

- Several challenges and trends threaten our ability to think critically—and make critical thinking more important than ever. These include biases, polarization, social media filter bubbles, fake news, and deepfakes.

- Anyone can train themselves to think critically. At a very basic level, you need to assess the information in front of you, look for gaps in the information, turn to trusted sources for additional information, ask lots of questions, and form your own opinions on topics.

One of the many benefits of critical thinking is it enables you to make better decisions. This leads us very neatly onto the next future skill: decision-making.

Notes

1. The Parties in Our Heads: Misperceptions about Party Composition and Their Consequences; *Journal of Politics*; https://www.journals.uchicago.edu/doi/abs/10.1086/697253

2. Exposure to opposing views on social media can increase political polarization; PNAS; https://www.pnas.org/content/115/37/9216.full

3. New Survey Reveals Teens Get Their News From Social Media and YouTube; Common Sense Media; https://www.commonsensemedia.org/about-us/news/press-releases/new-survey-reveals-teens-get-their-news-from-social-media-and-youtube

4. 27 Alarming Fake News Statistics on the Effects of False Reporting; Letter.ly; https://letter.ly/fake-news-statistics/

5. Immunizing the public against misinformation; WHO; https://www.who
.int/news-room/feature-stories/detail/immunizing-the-public-against-
misinformation

6. Even If It's 'Bonkers,' Poll Finds Many Believe QAnon And Other Conspiracy
Theories; NPR; https://text.npr.org/2020/12/30/951095644/even-if-its-
bonkers-poll-finds-many-believe-qanon-and-other-conspiracy-theories

7. State-sponsored disinformation in Western democracies is the elephant in the
room; Euronews; https://www.euronews.com/2020/07/06/state-sponsored-
disinformation-in-western-democracies-is-the-elephant-in-the-room-view

8. Former U.S. president Donald Trump launches TRUTH social media
platform; Reuters; https://www.reuters.com/world/us/former-us-president-
donald-trump-launches-new-social-media-platform-2021-10-21/

9. Facebook bans deepfake videos in run-up to US election; *Guardian*; https://
www.theguardian.com/technology/2020/jan/07/facebook-bans-deepfake-
videos-in-run-up-to-us-election

10. What are deepfakes and how can you spot them; *Guardian*; https://www
.theguardian.com/technology/2020/jan/13/what-are-deepfakes-and-how-
can-you-spot-them

11. Denying accuracy of Access Hollywood tape would be Trump's biggest lie;
Guardian; https://www.theguardian.com/us-news/2017/nov/29/denying-
accuracy-of-access-hollywood-tape-would-be-trumps-biggest-lie

CHAPTER 6
JUDGMENT AND COMPLEX DECISION-MAKING

You make hundreds, probably even thousands, of decisions each day. Many of these will be straightforward, fast decisions—which outfit to wear, which sandwich to have for lunch, and the like. But many decisions are not so easy. Indeed, I'd argue that decision-making is overall becoming harder (or at least more complex), in part because the world is now so fast paced, requiring us to think and act faster than ever, and in part because we have more information than ever before, making us prone to information overload. These factors complicate our ability to make good decisions—while also making judgment and decision-making skills more important than ever.

Understanding Judgment and Decision-Making

Decision-making simply refers to the process of making a decision. So what's the difference between judgment and decision-making? In essence, one underpins the other: judgment is the foundation of decision-making, since, when you make a conscious decision (as opposed to an automatic, instinctive response), you're judging one outcome or action as better than the other.

Defining judgment

Professor Andrew Likierman of the London Business School says, "Judgment is the ability to combine personal qualities with relevant knowledge and experience to form opinions and make decisions." I like this definition because it recognizes the role personal qualities (such as likes, values, and beliefs) play in judgment. After all, so many decisions facing us have no obvious right or wrong—and the "right" decision for you may be completely wrong for someone else. Things aren't always simple, in other words, and this is why we need to exercise judgment in order to arrive at the best decision we can, based on a combination of information, experience, and our personal attributes.

Likierman says judgment can be measured according to six core elements:[1]

1. What you take in (essentially, how much attention you pay to what you hear or read)

2. Who and what you trust (as in, whether you're basing judgment on high-quality, reliable raw materials; see Chapter 5, "Critical Thinking")

3. What you know (bearing in mind that what you know to be right today may be wrong or out of date tomorrow)

4. What you feel and believe (i.e. being aware of your values and beliefs, not only to use them when appropriate, but to also recognize when they may hamper judgment)

5. Your choice (as in, bringing together the raw materials and using decision-making techniques to improve the chances of success)

6. Delivery (in other words, making a choice isn't the end of the story, because you also have to take into account feasibility of delivery)

To further understand judgment and decision-making, we need to grasp two key concepts: rationality (or, more specifically, why humans aren't rational all the time), and intuition (or, importantly, the difference between acting on intuition and acting on considered logic).

Rationality

Good decisions are rational decisions, right? As humans, we all intend to make rational decisions. (No one intentionally sets out to make an irrational decision.) And this is why we do things like weigh up the pros and cons of decisions—a perfect example of a rational decision-making technique.

The trouble is, various limitations prevent us from being 100 percent rational all the time. Being pressed for time may limit a person's ability to gather all the information needed to make a fully rational decision. The limitations of human memory mean we can't retain infinite amounts of relevant information. And, of course, cognitive biases (see Chapter 5) limit our ability to think critically. This notion of *bounded rationality*—meaning that rationality is limited—was proposed by Nobel Prize–winning psychologist Herbert Simon. Simon's bounded rationality framework is so important because it explains the discrepancy between the assumed rationality of human decision-making and, well, the reality of human decision-making.

Another factor in decision-making is the way in which the brain actually *thinks*.

Intuition versus slow thinking

When we make a decision or take an action, the brain deploys one of two modes of thinking. Firstly, there's the quick, intuitive approach—where our brain processes a decision within seconds or milliseconds, often without any conscious thought process. Think of knocking over a full glass of water on your desk and then instantly snatching it up; your brain doesn't need to weigh up the pros and cons of picking up the glass versus letting the water spill towards your laptop; it just does it.

And then there's the second, more considered approach, where you think critically about the right way forward. A good example might be weighing up two job offers based on factors like salary, organizational fit, commute time, opportunities for future advancement, and so on.

This contrast between the two systems of thinking is perfectly described by renowned psychologist Daniel Kahneman in his best-selling book *Thinking Fast and Slow*. The first system, thinking fast (intuition), is perfectly fine for little decisions, such as deciding whether you want chicken or hummus in your sandwich. But for more complex decisions we need to deploy the second system, thinking slow. Here, we need to take the time to weigh up our options based on multiple factors. We need to look at the bigger picture, and not just rely on our instincts.

We all know this at heart. And yet every one of us has made rash decisions at one time or another—where we've rushed in and acted without proper consideration, only to regret it later on. We do this because thinking slow takes effort, concentration, and self-control—and, sometimes, maybe because we're tired or stressed, we just don't have those skills. As a result, the fast-thinking mode takes over. And this is precisely why even smart people can make bad decisions.

Bad decisions and mental shortcuts

History is littered with examples of otherwise intelligent people making terrible decisions. (Think of Ronald Wayne, co-founder of Apple, who sold his 10 percent stake in the company to co-founders Steve Jobs and Steve Wozniak for $800 in 1976—a stake that would be worth about $80 billion today. Wayne later said he "made the best decision with the information available to me at the time.")

One of the key reasons for bad decisions is simplification. When we're overwhelmed, we tend to oversimplify things. The brain, struggling to process a lot of data, creates mental shortcuts or simplifications in order to streamline the process and make what seems like a daunting decision (even if it's not really daunting but we're just tired or stressed), much simpler.

Overwhelm isn't the only reason why the brain creates a mental shortcut. You might, for instance, be trying to please someone you love or trying to be more like someone you admire, both of which involve the brain taking a mental shortcut ("What would make so-and-so happy?" or "What would so-and-so do in this situation?"), rather than taking the time to fully consider all the options.

Biases are another example of a mental shortcut. For example, there's the tendency to prioritize the information that supports what we already think (confirmation bias, discussed Chapter 5). Then there's the optimism bias, in which we tend to underestimate the likelihood of something bad happening to us. Take smoking as an example. Science clearly shows that smoking can kill. Most of us accept this as fact. But because of optimism bias, an individual smoker will believe smoking kills *other people*, but not them. As a result, they'll make the poor decision to carry on smoking.

Why Judgment and Complex Decision-Making Skills Matter More Than Ever

This tendency to oversimplify things and take mental shortcuts is a significant problem because, these days, we're constantly bombarded with information, especially online and on social media. And the more we're presented with overwhelming amounts of information—or even ambiguous information—the more we revert to these decision-making simplifications. We might, for example, boil a complex issue down into a simple "yes or no" decision (known as binary thinking), or allow biases to influence a decision. And we can do this even though we're trying hard to be rational.

Therefore, it's really important we all learn to recognize how our brain is processing information—as in, whether we're reverting to quick or slow thinking—and override the tendency to simplify issues. (Again, this is where critical thinking plays such an important role in decision-making, because it forces us to question whether we're seeing the full picture.) This isn't always easy in an age where we're bombarded with information all the time, but it is important.

We also can't ignore the fact that, as a result of AI, automation, and ever-growing quantities of data, machines will play a greater role in decision-making in the future, especially within organizations. Machines can, after all, analyze data with a speed, accuracy, and depth that humans could never replicate. This is great, but it doesn't replace human decision-making. In my mind, it only makes human decision-making even more important.

The final decision on what to do, based on what the data tells us, must come down to humans. Machines may be able to determine what appears to be the best way forward, based on the data, but they're unable to consider the wider implications of that decision—for example, the impact on company strategy, the people who work in the business, and the organizational culture. It's humans who will have to weigh up the broader impacts of decisions. This is why I believe decision-making will become more important, not less, in the digital age.

How to Improve Your Judgment and Decision-Making

Like all of the skills in this book, judgment and decision-making can be honed, practiced, and improved. And because we make countless decisions every day, we have endless opportunities to train ourselves to make better decisions.

Let's look at what individuals and organizations can do to promote better decision-making.

For individuals

Here are some practical ways to improve your decision-making skills.

- Start by defining the problem. Making a decision always starts with understanding the situation or problem. By doing this, you can identify gaps in your knowledge and where you need to gather extra information.

- Define your desired goal. What are you hoping will happen here? What's the best possible outcome? Sometimes we can get so bogged down in options that we forget what we're actually trying to accomplish.

- Outline and then weigh up your options. Making a list of options and then defining the pros and cons of each is a great way to make an informed, objective decision. Alternatively, you could use a SWOT (Strengths, Weaknesses, Opportunities, and Threats) analysis to evaluate different scenarios. If you have too many options, try limiting your choices by considering just a few options at a time.

- Get help when you need it. Depending on who is affected by the decision, you may need to involve others in the decision-making process. Other times, you may just need a second opinion, in which case talking things through with a trusted friend, colleague, or mentor will help you evaluate options, validate your decision, and grow in confidence.

- Set a deadline for decisions. If you're prone to procrastination, try setting a limit for how long you'll take to decide. Of course, some decisions really do need more time for consideration than others, so this isn't a one-size-fits-all approach. This leads to the next point.

- Keep the decision in perspective. Here, it can help to remember the *Thinking Fast and Slow* book (which I highly recommend you read). Not all decisions require slow thinking, and if you dedicate too much time and focus to the decisions that aren't really that important, it can sap your energy for making bigger decisions. And sometimes, it may help to weigh up the consequences of making a decision versus *not* making a decision—meaning that sometimes you can well afford to take a wait-and-see approach.

- Remember that judgment means drawing upon personal qualities as well as information and experience. In other words, good decision-making may involve a combination of head (information, experience), heart (values, beliefs), and gut (instincts). Being analytical is a good thing, but machines are always going to be better at that than humans—where we excel is in considering the very human consequences of our decisions.

- Be aware of your biases. While you shouldn't disregard gut instinct, do remember that, sometimes, gut instincts come from a place of personal biases. Always be on the lookout for this, and practice your critical thinking skills.

- Don't be afraid to experiment. More often than not, there is no one *right* decision—just different consequences. That guy who sold his Apple shares for $800? He has spoken openly about struggling to keep up with Jobs and Wozniak and how, if he'd stayed on in the company, he would have ended up "the richest man in the cemetery." The consequences may have been bad from a financial point of view, but that doesn't mean it was the wrong decision for him personally. With this in mind, don't be afraid to try different approaches or experiments to test certain decisions.

- Keep working at it. A good way to improve your decision-making is to keep analyzing decisions you've made and how they played out. Use that information to help you make similar decisions in future.

As a parent, I also think it's important to instill judgment and decision-making in the next generation. Some good ways to do this with your own children include:

- Let them practice making choices.

- Be clear about what they have decision-making control over, and in which areas you have parental control. Obviously, this will evolve as they grow older.

- Show them how you make decisions yourself, even tiny everyday decisions like whether to take a waterproof jacket when you walk the dog.

For organizations

It goes without saying that there are countless training options out there that will help you build decision-making skills in your organization, so let me focus here on the bigger issues for companies.

First, it's vital business leaders remember that, while machines will play a greater role in organizational decision-making, it's up to humans to consider the real-world consequences of decisions. This process of considering the impact of decisions on the business and the people who work for it should be modeled at every level of leadership and management.

As part of this, you'll want to build a culture of critical thinking, where people routinely do things like question data and consider biases. In such an organizational culture, people at all levels of the business are encouraged to ask questions and challenge decisions.

Key Takeaways

In this chapter, we've learned:

- Judgment is the foundation of decision-making; when you make a decision, you're judging one outcome or action as better than the other.

- Many decisions have no obvious right or wrong—and the "right" decision for you may be wrong for someone else. Remember, things aren't always simple, and this is why we need to exercise judgment to arrive at the best possible decision.

- Judgment means drawing upon personal qualities as well as information and experience. Therefore, good decision-making may involve a combination of head (information, experience), heart (values, beliefs), and gut (instincts).

- In this age of information overload, the ability to override mental shortcuts (such as binary thinking) and take considered decisions will be more important than ever. It's vital we all learn to recognize how our brain is processing information—quick versus slow thinking—and override the tendency to simplify issues.

- While machines will play a greater role in decision-making in future, it's up to humans to consider the real-world, human consequences of our decisions.

There's no doubt that emotion and intuition play a role in decision-making. So let's turn to the ability to interpret and understand emotions—otherwise known as emotional intelligence.

Note

1. The six elements of judgement; London Business School; https://www .london.edu/think/the-six-elements-of-judgement

CHAPTER 7
EMOTIONAL INTELLIGENCE AND EMPATHY

Many would argue emotional intelligence is more important and potentially a bigger predictor of success than general intelligence. But will that still be the case in the future, as our workplaces (and indeed our lives) evolve to center even more around machines and digital interactions? The very fact that I've included this chapter should give you a clue as to which way I lean. In my mind, for as long as there are humans in the workplace and human-to-human relationships, we will always need emotional intelligence and empathy.

What Is Emotional Intelligence and Empathy?

Emotional intelligence is the ability to be aware of, express, and control our emotions—and to understand and respond to the emotions of others. An emotionally intelligent person is aware that their emotions influence their behavior and impact those around them, and is able to manage those emotions accordingly, and even influence the emotions of others.

Empathy—or the ability to see the world from someone else's perspective—is a key component of emotional intelligence, since this gives us a great insight into how others are feeling.

EQ versus IQ

Emotional intelligence is also known as emotional quotient (EQ), recognizing the contrast between emotional intelligence and cognitive intelligence (IQ). Leading psychologist Daniel Goleman, author of the book *Emotional Intelligence*, is just one of the many experts who believe EQ is more important than the traditional predictor of success, IQ, because IQ is simply too narrow to represent the wide spectrum of human abilities.

While there may not be a definitive answer on which one is more important for success, it's certainly clear that IQ isn't the only predictor of success—EQ plays a huge role. As an example, at one insurance company, sales agents who ranked highly for EQ measures were found to sell insurance policies worth an average of $114,000—while colleagues who ranked lower for EQ measures sold policies with an average premium of $54,000.[1] That's quite a performance difference!

Notice I used the term "EQ measures" there, because EQ can be objectively measured, just as IQ can be measured. In fact, there are a number of different assessments to measure EQ, ranging from self-reporting tests to ability tests that are assessed by third parties. EQ can be measured according to the various levels or components of EQ, including:

- The ability to perceive emotions—through feelings, verbal language, and nonverbal signals such as body language

- The ability to apply emotion to cognitive tasks—such as thinking and problem-solving

- The ability to understand emotions—including the meaning behind those emotions

- The ability to manage emotions—the highest level of EQ (This involves regulating emotions so that we respond appropriately to situations and the emotions of others.)

Machines are getting better at detecting emotions

One fascinating area of EQ research is around detecting emotions. Typically, humans do this through verbal cues and nonverbal signals such as facial expressions and body language. But now, researchers are finding new ways for machines to detect emotions, for instance, by analyzing smells (machines are already learning to detect odors[2]), monitoring voice (AI has been used to accurately diagnose PTSD in veterans based on voice analysis[3]), and even using wireless signals bounced off the human body.

Let's dwell on that last one for a second, because it sounds pretty bonkers. But it's true. Scientists at Queen Mary University of London have developed a way to detect basic emotions such as anger and joy by emitting radio waves (like those emitted by Wi-Fi routers) towards participants and measuring the signals that bounced back. Incredibly, by analyzing changes in the signals bouncing back, caused by tiny body movements, the team was able to determine information about the participant's heart rate and breathing, which in turn provided basic insights into how that person was feeling.

Elsewhere, sensors and computer vision systems are being developed that analyze body posture, facial expressions, and gestures to detect human emotions.[4]

Imagine the implications of all this in the future, especially considering the plethora of smart wireless devices that are already in our homes. In theory, your smart lightbulbs and smart speaker could detect when you're feeling stressed or sad and adjust the lighting and music accordingly! Alternatively, sensors in your car could detect that you're feeling angry while driving, and take over more of the driving processes to ensure safety.

In fact, in the future, I believe machines will be *better* at detecting emotions than humans, in the way that machines are better at analyzing data than humans. Which begs the question, will humans even need EQ skills in future? (Spoiler: we will.)

Why We (Still) Need Emotional Intelligence and Empathy

EQ matters in all areas of life—work, relationships, you name it. Let's explore why that's the case.

The benefits of being emotionally intelligent

Among the many benefits, people with high levels of EQ are better equipped to:

- Form meaningful relationships—with friends, colleagues, romantic partners, family members, and so on—and sustain those relationships.

- Be more self-aware.

- Be a better listener.

- Think before they act.

- Better manage their emotions, especially in times of stress. Research shows that emotionally intelligent people are less stressed and anxious than others.[5]

- Diffuse negative emotions in others.

- Resolve conflicts.

- Have difficult conversations without hurting other people's feelings.

- Work collaboratively and perform well in interpersonal situations (for example, providing great customer service).

- Be a better leader. Because emotionally intelligent leaders are better attuned to the emotions of others, their team members feel seen and heard. On the flip side, research shows leaders who lack emotional intelligence cause higher employee turnover and lower employee engagement.[6]

- Make better decisions (see Chapter 6), because emotionally intelligent people recognize how they are feeling and how this may impact their actions.

EQ in the digital age

If machines might eventually be better at understanding emotions, will this replace the need for humans to be emotionally intelligent and empathetic? Absolutely not. While AI may be used for things like detecting and interpreting emotions in customers and employees, human-to-human interactions will still rely on EQ skills. For example, we need human EQ skills to create a positive experience for customers, or to ensure a pleasant, collaborative working environment.

I see AI as providing individuals and organizations with an EQ boost—augmenting our human-to-human interactions, or making machine-to-human interactions more engaging. A great example of using technology to improve interactions comes from Stanford University. Here researchers gave children with autism a pair of Google Glass smart glasses and a smartphone app that helped the children interpret facial expressions; after a few months of regular use, parents reported that their children made more eye contact and related better to others.[7]

Looking beyond interactions, complex decisions—which will still be made by humans, not machines (see Chapter 6)—also require EQ skills, since emotion is intrinsically linked to judgment and decision-making. (Not to mention the fact that we need empathy to help us understand the impact of decisions on others.)

At the very least, then, human EQ will remain as important as it is today. But you could argue that human EQ will become even *more* important in the digital age.

For one thing, EQ helps us slow down and take the time to consider things more fully, as opposed to letting our emotions dictate our behavior. This is important because we now live in a world in which we can get whatever we want—information, goods, services, even attention—within seconds, with just a few taps on a screen. What's more, we're constantly bombarded with new information and alerts that try to grab our attention. As a result, our attention spans are getting shorter and we may struggle to be alone with our thoughts. (Remember Dr. Anna Lembke's chilling warning from Chapter 4 that we're "losing our capacity to delay gratification, solve problems and deal with frustration and pain in its many forms.") Emotional intelligence can help us combat the negative effects of life online, by enabling us to be more present with our thoughts and feelings, and take the time to solve complex problems.

There's also the issue that digital tools may be making us less empathetic. One study that measured empathy in students over the course of 30 years found that empathy began to decline sharply after the year 2000, when digital tools began to creep in.[8] You need only venture onto Twitter or the average comments section on a tabloid newspaper website to see that many people are simply losing their ability to relate to others as human beings. If we're going to thrive in the digital age, we must retain our humanity—and EQ is a critical part of what makes us human.

Coping with digital transformation

It should be clear from the earlier chapters in this book that companies are entering a period of rapid transformation, brought about by new technologies. Implementing technology changes is one thing, but one of the hardest parts of digital transformation is managing the human aspect. Humans present some pretty big barriers to technological change—for example, by resisting change, or being cynical about new technologies. EQ helps business leaders overcome these barriers by empathizing with those in the business, listening to their concerns, communicating why change is necessary, and inspiring others to embrace new technologies.

EQ can also help combat some of the side effects of shifting to remote or hybrid work. When managers and team members are dispersed across different locations, it can make it harder to maintain social connections. (In one survey, conducted during the COVID-19 pandemic, almost half of newly remote workers reported that their sense of belonging had suffered.[9]) EQ can help to combat this by, at the very least, being more aware of the challenges others face when they work remotely, and empathizing with those challenges. Emotionally intelligent managers will also be better equipped to spot which employees are struggling and offer assistance in a supportive way.

Bottom line, in this age of rapid digital transformation and rising automation, I believe EQ will be more important than ever. Yes, machines will get better at a huge range of tasks, but one thing a machine will never be able to do better than a human is *relate to another human.*

How to Improve Your Emotional Intelligence and Empathy

EQ may seem like one of those qualities that you either have or don't have. But it's absolutely something you can learn and improve. At a very basic level, you're aiming to be more aware of your emotions and the emotions of those around you—and ideally, learning to regulate those emotions. Let's explore some simple ways to achieve this.

For individuals

- Listen to others. To understand how others are feeling, you first have to pay attention. What are they telling you, both with their words and their body language?

- Practice empathizing. I realize this tip sounds obvious, but think about it for a second—when did you last consciously put yourself in someone else's shoes? With our busy, stressful lives, we don't always take the time to stop and think "How would I feel if I was in their situation?" So try practicing this more often when you talk to people. You might be surprised at how powerful this exercise is when it comes to grasping another person's point of view.

- Identify and analyze your own emotions. Understanding your own emotions is a key part of EQ, so try to observe how you're feeling—perhaps even apply a label to the feeling, such as anger or happiness—and consider how this might influence your behavior and decisions. Over time, as you become more accustomed to analyzing how you feel, you should find it easier to stop and think before you act, rather than letting emotions take over.

- Be mindful. A great way to tune into your emotions is to practice mindfulness (being fully present in the moment). Take a few moments to really focus on everything that's going on inside and around you—thoughts, feelings, sensations, and the like—and use this understanding to analyze your feelings in more detail. This can be especially useful in times of stress.

You might also want to check out online tools to measure your EQ, and online courses designed to promote EQ.

For businesses

Assessments can be a useful tool to get people thinking about EQ and identifying areas for improvement, so I'd encourage businesses to invest in EQ assessments for leaders and teams. Building on this, there are numerous training programs to help people improve their EQ. I'd also encourage business leaders to think outside the box and consider other related learning opportunities. Mindfulness is a great example of a technique that can help people be more in touch with their emotions, and in turn boost their EQ.

Consider also the impact that increased remote working will have on EQ within the organization. Managers will need to take extra care to maintain social and emotional connections with remote team members—for example, through regular one-on-one video chats and group video meetings.

Finally, think about the future potential to build EQ into your digital systems—for example, a system that analyzes customer service calls in order to detect emotions such as frustration and anger.

Key Takeaways

To quickly recap what we've learned about emotional intelligence:

- Emotional intelligence (EQ) is the ability to be aware of, express, and control our emotions—and to understand and respond to the emotions of others. Empathy (the ability to see the world from someone else's perspective) is a key component of this.

- Machines are getting better at detecting emotions, and may even be better at it than humans in future. But human-to-human interactions will still rely on EQ. One thing a machine will never be able to do better than a human is *relate to another human.*

- EQ will become even *more* important in the digital age, helping us to slow down, be more present with our thoughts and feelings, and combat the negative side effects of life online.

- EQ can be learned and improved by practicing skills such as listening, empathizing, and analyzing your own emotions.

Let's turn to another area where humans—for now, at least—continue to outperform machines: creativity.

Notes

1. Is IQ or EQ more important?; Very Well Mind; https://www.verywellmind.com/iq-or-eq-which-one-is-more-important-2795287

2. AI is acquiring a sense of smell that can detect illnesses in human breath; The Conversation; https://theconversation.com/ai-is-acquiring-a-sense-of-smell-that-can-detect-illnesses-in-human-breath-97627

3. Speech-based markers for posttraumatic stress disorder in US veterans; *Depressions & Anxiety*; https://onlinelibrary.wiley.com/doi/abs/10.1002/da.22890

4. Sensors to Detect Human Emotions: The Newest Developments and Applications; Azo Sensors; https://www.azosensors.com/article.aspx?ArticleID=2238

5. Relationship between general intelligence, emotional intelligence, stress levels and stress reactivity; NCBI; https://www.ncbi.nlm.nih.gov/pmc/articles/PMC4117081/

6. Why emotional intelligence is important in leadership; Harvard Business School; https://online.hbs.edu/blog/post/emotional-intelligence-in-leadership

7. Google Glass helps kids with autism read facial expressions; Stanford Medicine; https://med.stanford.edu/news/all-news/2018/08/google-glass-helps-kids-with-autism-read-facial-expressions.html

8. Deconstructing Empathy in the Digital Age; Impakter; https://impakter.com/deconstructing-empathy-in-the-digital-age/

9. Remote work in the age of COVID-19; Slack; https://slack.com/blog/collaboration/report-remote-work-during-coronavirus

CHAPTER 8
CREATIVITY

Creativity has without doubt enabled us to become the species we are. And I'm not just talking creativity in terms of fine art, music, performing arts, architecture, and the like. I'm talking about a humbler definition of creativity: the ability to dream a different future and bring ideas to life. When you think of creativity this way, it becomes startlingly clear that we all have the ability to be creative, and most of us use that ability every day—whether you're an artist or an accountant.

What Is Creativity?

Creativity is the act of turning imaginative ideas into reality. Therefore, to be creative is to go through two processes: thinking and then producing. The producing part is key. You may have amazing ideas, but if you don't act on those ideas, you're not creating. In other words, imagination isn't the same as creativity; creativity means *bringing ideas to life*. Again, this doesn't mean you have to produce a museum-worthy piece of art to be deemed creative; producing solutions to a problem by imagining different scenarios is creative. Whatever the outcome, we're talking about bringing something new into the world.

In his book *The Creative Spark: How Imagination Made Humans Exceptional*, Augustin Fuentes argues that this ability to imagine something and then make it real is what differentiates us from other species—and has driven the evolution of the human race. From creating complex (and beautiful) tools to domesticating plants and animals, creativity has helped us feed ourselves and overcome challenges for hundreds of thousands of years. By seeing the world as it is, imagining how it could be, and then turning those ideas into actions, creativity has allowed our species to shape the world around us. And that is what makes us different.

Thinking of creativity in this way—as a fundamental human skill that we all have, and have had for as long as our species has existed—explodes the notion that creativity is something that only special, "gifted" people have (and are born with). To bust this myth, researchers at Exeter University in the UK studied outstanding performances in the arts, sports, and mathematics to discover whether excellence was determined by innate talent (i.e. from birth). The study concluded that excellence was achieved through a combination of training, motivation, encouragement, opportunities, and, most of all, practice.[1] To put it another way, even Mozart had to work his butt off before he created a masterpiece.

So the next time you hear someone say they're not creative, you can gently correct them. Because we all have the ability to create. It isn't a rare gift handed out to a chosen few at birth.

Why Does Creativity Matter?

Creativity will be a must-have ingredient for success in the fourth industrial revolution. Let's explore why.

The importance of creativity in the workplace

Creativity is often confused with innovation, but they're not strictly the same thing. Innovation is the process of creating value by introducing new goods and services, or improving existing goods and services. But since this isn't possible without creativity, I think it's fair to say that innovation and creativity are inextricably linked. Therefore, anyone who wants to be innovative in their job, or any organization that wants to be innovative, needs to foster creativity.

This is important to recognize because creativity and work aren't always seen as natural bedfellows. Creativity (rightly or wrongly) implies freedom and fun, while work implies, well, work: sitting at a desk or standing on a shop floor, or wherever, for eight hours a day. But creativity absolutely belongs in the workplace. It enables creative thinking (i.e. coming up with new ideas, imagining beyond the status quo), and problem-solving (implementing ideas to fix issues or make things better).

In this way, creativity makes use of skills that are highly attractive in the workplace, including:

- Critical thinking (see Chapter 5)—because before you can imagine new possibilities, you first need a clear understanding of the current situation

- Open-mindedness—because imagining new solutions means you have to let go of the "this is how we've always done things" mentality

- Judgment and decision-making (see Chapter 6)—because you may have to weigh up the pros and cons of different future scenarios and outcomes before you can decide on the best way forward

- Courage—because trying new things often means taking a risk, however calculated

- Communication and collaboration—because bringing new ideas to life generally requires the help of others (We'll delve into collaboration in the next chapter, and communication in Chapter 10.)

Given this connection between creativity and other desirable attributes—not to mention innovation—it's hardly surprising that the World Economic Forum listed creativity as one of the 10 essential skills needed to thrive in the fourth industrial revolution (in any profession or industry, not just creative ones); in fact, creativity ranked third, just behind problem-solving and critical thinking.[2]

In another report, 58 percent of employers said they expected creativity to grow in importance in the coming years.[3] Why? It's because the nature of work itself is changing.

The rise of machines makes creativity more important than ever

In his book *A Whole New Mind: Why Right-Brainers Will Rule the Future*, Daniel Pink argues that the sort of left-brain linear thinking that's prized in many of today's workplaces will ultimately be replaced by right-brain skills such as creativity and empathy. These right-brain skills will drive a new age of economic development that Pink refers to as the Conceptual Age (which follows the Information Age, which has been driven by knowledge workers, and the Industrial Age before that, which was driven by factory workers, and the Agriculture Age before that, which was driven by farmers). In Pink's vision, creativity will drive competitive advantage. The future will belong to creative thinkers, he says.

What's behind this bold claim? Quite simply, because the "knowledge work" that has powered economic growth in recent decades is increasingly being done by machines. And as more and more knowledge tasks are automated, human workers will transition into more thoughtful, innovative, creative work.

Machines will even enhance our creativity

We think of creativity as a uniquely human skill. But you might be surprised to learn that AI is getting better at stereotypically creative tasks like composing music and writing novels. For example, an AI algorithm created an artwork called *Portrait of Edmond de Belamy* that was later sold by Christies for a whopping $432,500.[4] To create the portrait, the AI was fed a data set of 15,000 portraits spanning six centuries. In another example, choreographer Wayne McGregor used AI to suggest choreography options based on hundreds of hours of video footage of dancers.[5]

What you'll notice from these examples is that the AI learns from other (human) artists' works and then uses those models to suggest or create new versions. This is certainly impressive, but it doesn't replace, or even equate to, human creativity. An AI can compose a piece of music in the style of Mozart, but it can't come up with a completely new model for composing. Machine creativity relies on human creativity for the underlying models. Therefore, it's humans—not machines—who will continue to drive the creative process and push the boundaries of creativity.

That said, I do believe machines will play a greater role in *aiding* the creative process, for example, by suggesting new product designs based on parameters and specifications set by a human designer. This is precisely what happened when designer Philippe Starck worked with AI to design a new chair that was unveiled at Milan Design Week.[6]

This notion of humans working alongside AI to create something new is known as *co-creativity*, and I believe we'll see a lot more co-creativity in the future. As Professor Marcus du Sautoy, author of *The Creativity Code*, puts it, the role of AI will be to act as a "catalyst to push our human creativity." Therefore, creative skills must, in my view, go hand in hand with digital literacy skills (see Chapter 1).

Will machines ever truly be creative, without human intervention? Given that we've barely scratched the surface of what AI is capable of, it's certainly possible. But I believe the biggest potential lies in enhancing the creative work that humans do—whether it's creating an artwork, designing a new product, coming up with a marketing campaign, or solving customers' biggest problems.

Instilling creativity at a young age

I'll move on to boosting your own creative skills next, but first let me briefly discuss the importance of creativity in education. Given that creativity will be a vital skill for success in the future, I believe schools should be placing as much emphasis on artistic subjects as they do on subjects like science and math. I'm a governor at a local school, and I was surprised to find that the school was struggling to attract students to the arts subjects on offer at the school, probably because so much emphasis has been placed on the importance of STEM (science, technology, engineering, and math) subjects for future success.

Therefore, schools (and parents, for that matter) have much work to do if they're to sell creativity as an important skill. One way to do this, in my opinion, is to highlight the link between creativity and other, more traditional topics, such as technology and math. Technology, for example, is all about coming up with new ideas and solutions.

So instead of talking about the importance of STEM education, we should be talking about STEAM education (the A stands for arts). And yes, I know that creativity isn't just about the arts, but by creating opportunities for young people to be artistic, we lay the groundwork for the types of creative thinking and creative problem-solving that are necessary for success. (This is backed up by one interesting study that showed nearly all Nobel laureates in the sciences practice some form of art as adults, such as painting or singing.[7])

It's vital, then, that schools and parents begin selling the importance of creativity, and preparing young people for the kinds of jobs that will exist in the fourth industrial revolution—those right-brain jobs, as Daniel Pink puts it, rather than the left-brain jobs that can be performed easily by intelligent machines.

How to Boost Your Creativity

Bearing in mind that creativity isn't some divine gift handed down at birth, what practical steps can individuals and organizations take to enhance creative skills?

For individuals

Some of my favorite ways to boost creativity include:

- Ask questions. Deploy your critical thinking skills (Chapter 5) and ask questions that challenge how things currently are, so that you can imagine how they *could* be.

- Build your network. Hanging out with people who have different life experiences and knowledge is a great way to expose yourself to new ideas. So meet new people, have lots of interesting conversations, and watch your creative horizons expand.

- Go to new places. Travel is another great way to expand your mind. But this doesn't necessarily mean taking a big trip abroad. Something as simple as working from a different coffee shop or going on a new walk can boost your observation skills.

- Look for patterns. Richard Branson's mantra for success is ABCD—Always Be Connecting the Dots—so try to draw connections between different questions, ideas, and problems. This may highlight interesting solutions.

- Keep a journal. I keep an ideas journal handy and I'm always jotting things down—an interesting theory from an article I've read, a new phrase I haven't heard before, a new book idea, interesting observations, anything. Doing this, I've noticed I make surprising connections between unexpected topics.

- Read more. I know it can be hard to find time for this in today's busy world, which is why I like to listen to audiobooks while I'm running. This often sparks new ideas—which I'll capture right then and there as a voice note on my phone and then later transfer to my ideas journal.

- Unplug and create headspace. Having just said I like to listen to audiobooks while I run, I also make sure to regularly run or walk without listening to anything at all! Being alone with your thoughts creates space for new ideas to bubble to the surface—and some of my best ideas have come during these periods of quiet.

- Daydream. From big daydreams like imagining a whole new career or small daydreams like visualizing the perfect anniversary meal, it's important to take time to dream. This can be tricky in this age of information overload, but you can boost your daydreaming capabilities by simply not reaching for your phone during idle moments.

- Maintain a positive mindset. Research shows that a good mood boosts the brain's ability to come up with innovative, creative thoughts, while a bad mood leads to more analytical, straightforward thinking.[8] So do practical things to keep your mood up, such as exercising, being mindful, expressing gratitude, and getting a good night's sleep.

- Practice, practice, practice. Tuning into your creativity is like learning to play the piano or run a marathon—it takes time and regular practice. So try to build some or all of the above steps into your regular routine.

For organizations

It's vital organizations create an environment in which people feel able to create. Google, for example, has famously told its teams that at least 20 percent of their time should be spent exploring or working on projects that won't pay any immediate dividends—but may lead to bigger opportunities in the future. People are free to spend a fifth of their time working on whatever they think may best benefit Google *at some point in the future,* while recognizing that this creative time may, ultimately, lead to nothing. The company credits this "20-percent rule" for delivering many of its biggest advances.[9]

This sounds great, but we have to recognize that it's difficult for people to create when they're bogged down in the minutia of everyday work and life. Science tells us as much. For example, research has shown that primates are more creative when their basic needs for food are taken care of. Conversely, people with low incomes have been shown to perform worse in problem-solving tests when reminded of their financial worries.[10] In other words, creativity thrives when the basics are taken care of. So how can you take care of the basics and allow people the space to be creative? In a workplace context, this may mean using technology and automation to cover more of the mundane tasks, thereby freeing up people's time and energy for more valuable work.

Looking at the bigger picture, people need to know it's okay to be creative and take risks, even if that ultimately leads to failure. As part of this, you'll want to build a culture that recognizes and celebrates creativity—which may mean rethinking things like performance management and metrics.

Key Takeaways

In this chapter we've learned:

- Creativity means turning imaginative ideas into reality. This involves two processes—thinking and producing. Thinking without producing is imaginative, but it's not creativity.

- As machines take on more of the "knowledge work," creativity will become an increasingly prized attribute in the workplace.

- Machines are getting better at some creative tasks, but cannot create in the way humans can. Regardless of the impressive capabilities of AI, it'll be up to humans to imagine future scenarios, dream up new possibilities, and steer the organizations of the future.

- Creativity isn't an innate gift we're born with. Anyone can be creative and practice creative skills through habits like daydreaming, embracing quiet, reading, having interesting conversations with new people, and so on.

Creativity has enabled our species to shape our world, but this wouldn't have been possible without collaboration. After all, bringing ideas to fruition generally requires us to communicate and engage with others, and enlist their help to turn ideas into reality. Therefore, creativity is more of a social process than you might think. Which brings us neatly onto our next vital future skill: collaboration.

Notes

1. What is Creativity?; Creativity at Work; https://www.creativityatwork.com/what-is-creativity/

2. The 10 skills you need to thrive in the fourth industrial revolution; World Economic Forum; https://www.weforum.org/agenda/2016/01/the-10-skills-you-need-to-thrive-in-the-fourth-industrial-revolution/

3. Why Is Creativity Important To Employers?; Education Scotland; https://education.gov.scot/nih/Documents/Creativity/CRE24_Infographics/cre24-why-is-creativity-important-to-employers.pdf

4. Is artificial intelligence set to become art's next medium?; Christie's; https://www.christies.com/features/A-collaboration-between-two-artists-one-human-one-a-machine-9332-1.aspx

5. Can Machines And Artificial Intelligence Be Creative?; *Forbes*; https://www.forbes.com/sites/bernardmarr/2020/02/28/can-machines-and-artificial-intelligence-be-creative/?sh=5f4644545803

6. AI rejects conservative views on furniture, designs wacky chair; The Next Web; https://thenextweb.com/news/ai-rejects-conservative-human-views-on-furniture-designs-wacky-chair

7. Arts Foster Scientific Success; *Psychology Today*; https://www.psychologytoday.com/files/attachments/1035/arts-foster-scientific-success.pdf

8. 3 science-based strategies to increase your creativity; TED; https://ideas.ted.com/3-science-based-strategies-to-increase-your-creativity/

9. Google Says it Still Uses the '20-percent rule' And You Should Totally Copy It; Inc; https://www.inc.com/bill-murphy-jr/google-says-it-still-uses-20-percent-rule-you-should-totally-copy-it.html

10. Where Creativity Comes From; *Scientific American*; https://www.scientificamerican.com/article/where-creativity-comes-from/

CHAPTER 9
COLLABORATION AND WORKING IN TEAMS

The fourth industrial revolution is going to present many challenges for organizations, not least keeping up with the breakneck pace of change. So it makes sense that businesses will want people on their teams who can work well with others to overcome challenges and drive the company forward. This is why collaboration is another of my top future skills. It's one of those skills that seems obvious for workplace success, and yet so many of us have encountered teams and individuals that just don't play well with others. In this chapter, we'll explore what makes a good collaborator, what prevents people from collaborating with others, and what collaboration may look like in the workplaces of the future, where remote, distributed teams will become the norm.

What Is Collaboration?

Collaboration means working with others to make collective decisions and achieve a common goal.

Are collaboration and teamwork the same thing?

Not exactly, although they both involve working together. A team is made up of individuals—with each individual being responsible for their own defined role and tasks, which contribute to the team's overall objectives.

Typically, the team will have a leader who oversees each individual's work and drives the team forward. Think of a soccer team, headed up by a coach, with each player fulfilling roles such as goalkeeper, defender, center forward, and so on. If one player is sent off the pitch, or if the coach isn't there to decide tactics, the team will struggle to function as well.

Collaboration involves teamwork, since it requires people to work together. But above and beyond that, collaboration means thinking together, making decisions together and sharing responsibilities—as opposed to working as separate individuals. There may not be a leader at all; the group could be a self-managed unit. And if one person isn't there, the rest of the group steps in to pick up the slack and continue towards their goal. That's collaboration.

With both teamwork and collaboration, the end result may be the same—meaning that the group achieves their desired outcome. However, the group dynamics may differ.

Think of it this way: a team within an organization can, in theory, fulfill its objectives even if the individuals don't especially like, respect, or trust one other—even if the individuals don't possess important skills like empathy and emotional intelligence. So long as everyone delivers on their responsibilities, the team can still meet the collective goal and be considered "successful." But that's not the same as collaboration. Working together in a truly collaborative way relies on things like emotional intelligence, mutual respect, and trust.

This distinction between teamwork and collaboration is important, given the changing nature of teams. Traditional top-down organizational structures are giving way to flatter organizational structures, organized around project teams rather than siloes and layers of management. Therefore, the teams of the future will need to be more collaborative than ever.

So what makes a good collaborator?

Perhaps one of the reasons why collaboration can be challenging is that it requires many different interpersonal skills. Obviously, communication skills are central to collaboration, and I'll talk more about interpersonal communication in Chapter 10. There's also emotional intelligence and empathy, which I talked about in Chapter 7.

Looking at other important qualities, good collaborators:

- Are active listeners. Rather than waiting impatiently for others to finish speaking so they can offer their two cents, good collaborators listen attentively to what people are saying.

- They're generous with their time, knowledge, experience, and encouragement. This can be especially challenging in today's super-busy world. I know I'm guilty of occasionally being too selfish with my time and not giving it as freely as I could.

- They're adaptable. Good collaborators possess that ability to be flexible and go with the flow when things don't go according to plan—as they often don't.

- They're trusting and trustworthy. Good collaborators create a safe space where people can share their ideas without fear. They're transparent and authentic. They deliver on their promises. All of this inspires trust. What's more, they give their trust freely, without skepticism and negativity.

- They're motivated. As I said, collaboration doesn't necessarily require a leader to set the agenda and hold people accountable. Good collaborators are self-motivated. They have an inner drive that propels them forward—and inspires others along the way.

- They're respectful. Because collaboration thrives in an environment of mutual respect.

- They're team-oriented. Good collaborators care more about achieving the common goal than being recognized for their individual role. They don't have a huge ego, in other words.

- They're open to feedback. Good collaborators don't get defensive when presented with feedback; they see feedback as an opportunity to learn.

Why Does Collaboration Matter?

Working well with others is important for almost all jobs. The advantages for businesses are clear. Collaboration allows individuals to work more efficiently, solve problems more creatively, be more innovative, and be more productive, which in turn drives business success. Plus, working collaboratively helps individuals and teams build better relationships, which in turn can boost factors like employee satisfaction, motivation, engagement, employer brand, and so on.

And for you as an individual, collaborating with others is almost certainly more efficient than going it alone and trying to do everything yourself! It also presents more opportunities to learn from others, potentially people from very different backgrounds, and to gain interesting new perspectives. To put it in corporate speak, knowledge transfer is enhanced in collaborative teams—and this benefits you in your current role, and your future career prospects.

Collaboration in the 21st-century workplace

Earlier in the chapter, I mentioned the trend for flatter organizational structures. (If you're interested, you can read more about this and other future trends in my book *Business Trends in Practice: The 25+ Trends That Are Redefining Organizations*.)

Another trend that will heighten the need for collaboration is remote and hybrid working, where team members may be spread across many different locations, and potentially across different countries. In the wake of the COVID-19 pandemic, 84 percent of employers said they were set to expand remote working—with the potential to move 44 percent of their workforce to remote working.[1] What's more, people will increasingly be working as "gig" workers, contractors, or freelancers. Therefore, the teams of the future will likely include a mixture of office workers, remote workers, contractors, permanent team members, and potentially other employees from within the business who "float" between projects and teams.

With such distributed teams, there's even more of a need for people to feel connected to one another and to a shared purpose. When this connection falters, people can genuinely suffer (see the impact on newly remote workers in Chapter 7). I'll talk more about boosting collaboration in remote teams later in the chapter, but suffice to say that collaboration will be more important than ever. Granted, collaboration may look a little different in future, and will rely more on digital tools—but it will still be a vital part of success, for individuals, for teams, and for entire organizations.

What's stopping good collaboration?

Working with others to achieve a common goal. Sounds simple, doesn't it? And yet it's not always that simple—otherwise all organizations would be shining beacons of collaborative harmony. Many of us know from painful experience that there are plenty of organizations, teams, and individuals out there that are anything *but* collaborative. Why? Aside from lacking the interpersonal skills I mentioned earlier in the chapter, it's probably because certain barriers are preventing good collaboration. These barriers may include:

- A culture of fear. When people are afraid of voicing new ideas, asking questions, or reaching out to others for help, collaboration—and innovation, for that matter—suffers.

- Time. If someone thinks it'll take too much of their time to collaborate with others, they'll probably just try and go it alone. It's a bit like creativity in that sense (see Chapter 8); people are more creative when they have the space and time to be creative, and when they aren't overwhelmed with basic tasks. The same can be said for collaboration.

- Flexible and remote working. Because, if we're honest, it takes more effort to collaborate with others when they aren't sitting in the same room.

- Poor leadership. Collaboration must be modeled at every level of the organization, so if a business has leaders who fail to inspire trust, or aren't generous with their time and knowledge, or don't listen to others, this will filter down into teams.

- Performance management systems. Many performance management systems actively pit employees against each other through targets and rewards. Metrics should recognize the importance of relationships, not just tasks.

- Assumptions about different personality types. Much has been written about the differences between introverts and extroverts, and how they behave in group settings. While I think it's important to recognize and celebrate differences, it shouldn't be an excuse for poor collaboration. To put it another way, just because someone is an extrovert, doesn't mean they aren't a great listener or aren't generous with their knowledge; and just because someone is an introvert doesn't mean they aren't emotionally intelligent or empathetic. It's important to focus on the specific qualities that make people good collaborators, rather than broad personality types.

How to Boost Your Collaborative Skills

While some people may lean more naturally towards collaboration than others, anyone can learn to be a better collaborator. Let's explore some of my favorite ways to boost collaborative skills.

For individuals

- Practice active listening. When others are talking, focus on what they're saying rather than formulating your own response in your head. This way, you'll be better able to understand their perspective, and they'll feel heard. To really nail this, you'll need to filter out distractions—by which I mostly mean don't look at your phone while others are talking!

- Volunteer your time and talent. If you aren't presented with as many opportunities to be collaborative as you'd like, go looking for opportunities yourself. Offer to get involved in projects (big or small) at work. Sign up for committees inside your organization, as well as industry groups. Volunteer to be a mentor. Some of the best learning opportunities may come from things that are outside of your everyday work.

- Find a mentor of your own. If you admire someone else's collaboration skills, ask if they'd be willing to mentor you. This could be as informal as grabbing a coffee once a month.

- Be open about what you need from others. Not everyone works in the same way, so say up front how you like to collaborate and what you need from others in order to collaborate well. In turn, ask what they need from you. And do deliver on what you promise.

- Hone your emotional intelligence and empathy skills (see Chapter 8).

If you're a remote worker, here are some extra ways to foster collaboration from a distance:

- Ask people's communication preferences. Some people prefer email, others like to jump on a telephone or video call, some people respond well to instant messaging and emojis, while others are immediately turned off by a yellow smiley face! Learn what your collaborators like, and use the right channels for the right people.

- Brush up on your communication skills. When you're not in the same room as your colleagues, they'll be less able to pick up on non-verbal clues, so what you say matters more than ever. Think ahead about what you want to say in meetings and one-to-one calls. And take time over your written communications to make sure they're clear. (On the flip side, try not to read subtext or tone into written communications that isn't really there.) Read more about communication in the next chapter.

- Shoot the breeze. To build and maintain rapport, make time for more casual conversations, of the sort you'd have in the office corridor or kitchen. You may even want to ask your boss if your team could have an instant messaging or chat channel that's purely used for informal chats, as opposed to work updates.

- Keep time differences in mind. If your colleagues are across different time zones, be respectful of that. Don't expect responses at a time when they're not supposed to be in the office, and ensure any project deadlines take the time difference into account.

- Try to meet up in the real world, when possible. Getting together offline can do wonders for building rapport. When meeting in person isn't possible, you could try organizing a video happy hour hangout, or virtual quiz.

For organizations

I can't stress enough how important it is to lead by example. Leaders and managers must be good listeners. They must be respectful of other's ideas and feedback. They must inspire trust, and be flexible, and all the things that make a person a good collaborator. The overall organizational culture must be one that prioritizes open-mindedness, transparency, and interpersonal relationships—which may mean you need to rethink or tweak your performance management metrics.

Of course, there are specific training and team-building activities that promise to foster collaboration, from off-site retreats to team-building exercises and games. I know some team-building activities can be a bit cringeworthy, but time spent working together as a team to achieve a goal—even if it's something silly, like building a LEGO model—can be great for strengthening collaboration skills and relationships.

Organizations will have to work hard to enable collaboration across remote teams. This means investing in the technology tools that make collaborating from a distance easier, such as document sharing platforms, online communication tools, video calling software, and project management software. Managers will also want to schedule regular check-ins with individuals, and arrange more social hangouts (whether virtual or in person) to build relationships within the team.

Key Takeaways

To briefly revisit the key takeaways on collaboration and working in teams:

- Collaboration means working with others to make collective decisions and achieve a common goal. This isn't strictly the same thing as teamwork, although both involve working together. While a team can be "successful" purely by everyone in the team fulfilling their defined responsibilities, collaboration requires more from people—it requires things like respect, trust, transparency, listening, and emotional intelligence.

- In the future, as organizations transition to flatter, more project-based organizational structures, and as more people work from home, collaboration will be more important than ever. That said, individuals and teams may have to work harder to foster collaboration when they're not located in the same place.

- There are many barriers to collaboration, including organizational culture, fear, lack of time, and performance management structures that prioritize tasks over relationships.

- Anyone can learn to be a better collaborator. Good ways to enhance your collaborative skills include active listening, volunteering your time and knowledge, being open about what you need to be a good collaborator, and working with a mentor.

As communication and collaboration are inextricably linked, let's dwell a little longer on the subject of interpersonal communication.

Note

1. The Future of Jobs Report 2020; World Economic Forum; http://www3 .weforum.org/docs/WEF_Future_of_Jobs_2020.pdf

CHAPTER 10
INTERPERSONAL COMMUNICATION

Message a company online these days and there's a pretty good chance you'll be interacting with a chatbot. Thanks to machine learning, these bots can now communicate with humans well enough that it's not always obvious whether you're talking to a real-life person or a machine. But despite these advances, it'll be a long time before machines can communicate as well as humans. Let's explore the field of interpersonal communication—and how these vital skills may need to adapt for the future of work.

What Is Interpersonal Communication?

Interpersonal communication is simply the exchange of information, emotions, and meaning between people. In the context of the workplace, this takes place constantly, in meetings and presentations, in emails, in those little watercooler chats—you name it.

Interpersonal communication spans four types of communication:

- Oral communication, which covers any form of spoken communication, including phone and video calls. Here, lots of factors inform how well you communicate your message, from your choice of words and tone of voice, to the speed at which you talk and the use of filler words like "um" and "ah."

- Written communication, which covers any form of written communication, such as emails, Slack messages, social media comments, reports, and so on. This involves writing clearly (so good grammar is important), and deploying symbols such as emojis to make your meaning clear.

- Nonverbal communication, or conveying meaning via physical cues such as body language and eye contact. (More on this coming up next.)

- Listening, which is a vital, and often overlooked, part of communication. To listen well is to listen *actively*, by which I mean giving the person speaking your full attention, and using encouraging signals such as eye contact and "uh-huh" type sounds. But listening also means being able to interpret what the person speaking really means, via their nonverbal signals, and using emotional intelligence (see Chapter 7).

It's not just what you say . . .

It's also how you say it, because it turns out that vocal and nonverbal cues like body language, eye contact, facial expressions, tone of voice, head movements, and hand gestures speak volumes. In fact, the overwhelming majority—as much as 93 percent—of communication is nonverbal.

Very often, these nonverbal signals are delivered and understood without us even realizing it, because our ability to communicate nonverbally is instinctual and innate, and as such often matters more than words themselves. In one study, participants' responses to spoken words were more dependent on the tone of voice than the connotation behind the word itself, so even a negative word like "terrible" was perceived neutrally or positively depending on how the word was delivered.[1] (If you've ever had someone tell you they were "fine," when their entire demeanor screamed that they were very much not fine, I'm sure you can relate.) This led the author of the study to conclude that the interpretation of a message is only 7 percent verbal, 38 percent vocal, and 55 percent visual, thus giving us the oft-quoted figure that nonverbal communication makes up 93 percent of understanding.

This is important because the ways in which we communicate have shifted drastically to written communication—emails, Slack chats, social media comments, and so on—in which these precious nonverbal signals are lost. Have you ever misconstrued someone's comment online, mistaking sarcasm or a joke for rudeness? It's because the nonverbal subtleties have been lost. Chances are, if it had been a face-to-face encounter, you would have gotten the irreverent meaning behind the comment loud and clear.

Luckily, more and more remote meetings are taking place via video platforms, which allows us to pick up on the nonverbal side of communication, as well as the verbal. I talk more about communicating remotely later in the chapter.

What type of communicator are you?

We don't all communicate in the same way. Broadly speaking, there are five different communication styles. You may not fall exactly into one category; you may recognize elements of two or more categories in yourself, or revert to a certain style in certain situations. The key thing is, if you can recognize your own communication style(s), and the styles of others, you'll be better placed to tailor your communication style to match different audiences.

The five communication styles are:

1. Assertive communication style
   Generally considered to be the most effective communication style, this means communicating confidently and calmly, clearly expressing your needs, but without riding roughshod over others in the conversation. Assertive communicators tend to use a lot of active "I" statements, such as "I feel frustrated when you show up late for our meetings" as opposed to less active statements such as "You need to stop turning up late." They communicate without emotional manipulation; rather, they consider and respect the rights and opinions of others, while standing up for their own needs, expectations, and boundaries.

If you want to be more assertive in your communications, inject confidence into what you say (even if you don't necessarily feel confident). Replace woolly words such as "could" "may," and "might" with "will." Listen actively to what others are saying and respond with clarity and calmness, while looking for mutually beneficial solutions.

To communicate well with an assertive communicator, ask them what they think and give them the space to share their ideas. They are solutions-oriented people, so they respond well to suggestions—and even criticism, providing it's delivered with respect.

2. Aggressive communication style
   While the assertive communicator seeks solutions and values what others have to say, the aggressive communicator wants nothing more than to get their own way. As such, they are frequently hostile, demanding and intimidating in their communications, and act like what they have to say is more important than everyone else in the room. Even if what they're saying is correct, the way in which they speak often turns people off.

   If you want to get on with others, this isn't a style that I'd recommend you deploy at work. But what should you do when you encounter an aggressive communicator? First things first, be prepared for the fact that they might try to steamroller you or communicate in a way you find off-putting. Then get to the point of the conversation as quickly as possible, keeping it short and sweet.

3. Passive communication style
   These communicators are people-pleasers, plain and simple. They're easygoing and out to avoid conflict at all costs, even if it means they don't convey their own opinions and needs (which may eventually lead to resentment). Passive communicators are happy to let others take the floor, and go along with the ideas of others, and as a result their contributions may not be heard.

   If this sounds like you, try to inject a little more confidence into your communications and emulate the assertive style (even if you're faking it at first). Remember, your contributions matter. Practice setting boundaries and saying no to unreasonable requests, so you don't end up overloaded and resentful.

To communicate effectively with a passive communicator, give them the opportunity to be heard by directly asking them what they think. Never dismiss their ideas outright or get confrontational, as it will dent their confidence; instead, keep things positive.

4. Passive-aggressive communication style
   A hodgepodge of passive and aggressive, this type of communicator may appear passive and easygoing on the surface, but underneath lies aggression, frustration, and resentment. As a result, they may go too hard on the sarcasm or come off as patronizing.

   Again, this isn't a style you want to be deploying in the workplace. But you'll probably come across passive-aggressive types at some point; in which case, don't be tempted to respond in the same style. Model an assertive, positive tone instead. Also, look for clues on why they might be communicating this way; for example, does their passive-aggressive style come out only at certain times, such as when they're stressed?

5. Manipulative communication style
   This type of communicator rarely says what they mean, and instead uses lies, manipulation, and emotional arguments to get the outcome they want. They may come off as two-faced, fake, or even patronizing. The manipulative communicator knows exactly what they want—just like the assertive communicator—but unlike the assertive communicator, they don't express it directly.

   If you find yourself falling into this style, try to emulate the assertive communicator and be more direct about what you want—while recognizing that your needs are no more important than anyone else's in the conversation.

   To communicate effectively with a manipulative communicator, try redirecting them away from emotional arguments and towards facts. Calmly but firmly sticking to your guns is a good way to show that you don't respond well to manipulation.

You can read more tips for boosting your interpersonal skills later in the chapter.

The importance of storytelling in communication

Twenty years ago I saw a presentation that forever changed how I give presentations. They delivered a speech with PowerPoint slides, yet every slide had nothing but pictures on it. No words—maybe the occasional number, but otherwise just pictures. The speaker then told a story around each slide, using anecdotes and real-world examples to give context. I was so impressed, I copied their approach, and continue to give presentations in the same way. It's revolutionized a part of my job that used to make me really nervous.

This was probably when I first learned the importance of storytelling in communication. By incorporating stories into your communications— especially presentations—you engage people and help them better understand your message. Storytelling is especially important when you're talking about numbers (see Chapter 2, "Data Literacy"). Numbers alone aren't engaging or memorable. Stories are. Which is why it's better to craft a narrative than spew data at people. For example, when I talk about the proliferation of smart, connected devices in our homes, I could reel off any number of statistics about the projected number of smart devices by 2030. Or I could paint a descriptive picture of the smart homes of the future, with smart lighting and speakers that use sensors to gauge your mood and adjust the atmosphere accordingly, and smart fridges that can tell when you're running low on ingredients and automatically order a grocery delivery for you. Which is more interesting?

Companies employ this storytelling strategy all the time. They tell stories of their history and values in order to engage employees and inspire customer loyalty. Politicians tell stories about themselves and their backgrounds in order to create an emotional connection with voters. Take a leaf out of their book and build storytelling into your own communications.

Why Do Interpersonal Communication Skills Matter?

This is one of those no-brainer skills that I probably don't have to sell you on. We all know that communication is important for workplace (and life) success. But let's briefly explore some of the biggest incentives to sharpen up your communication skills:

- Demand for people with interpersonal skills has been on the rise for decades. Harvard research shows that, over a 30-year period, jobs that require a lot of social interaction grew by 12 percentage points, while the number of not-so-social jobs shrank.[2]

- Interpersonal communication helps us to persuade others, negotiate, and resolve conflicts more effectively. In this way, communication skills are essential for collaboration (see Chapter 9).

- Communication is also linked to creativity. As we saw in Chapter 8, turning your ideas into action often requires the help of others. Great communicators are not just clear and precise, they're seen as more likeable—and this makes it easier to bring people on board with ideas and bring projects to life.

- Communication helps to build trust and relationships, especially when it comes to the more instinctual nonverbal communication. Think about it, if you're with a potential client and your nonverbal signals show that you're bored, distracted, and barely listening, how likely are you to close that deal?

- Finally, communication skills can't be outsourced to machines, or at least not to the same extent as other tasks and jobs. Interpersonal communication is unique to humans (the clue's in the name). And while chatbots may be able to handle straightforward communications with ease, they've got nothing on our ability to communicate in a lively, engaging, and, above all, *human* way.

Bottom line, employers can't afford *not* to have great communicators on their teams, and we as individuals can't afford to let communication skills slide. This is especially important to keep in mind as more and more of us work from home. When you have less "face time" with others, it can be easy to neglect those all-important, relationship-building communications, such as watercooler moments or giving a colleague encouraging nonverbal signals during a presentation they've worked hard on. We must all work hard to give interpersonal communication the attention it deserves.

How to Boost Your Interpersonal Communication Skills

Let's explore some of my favorite tips for sharpening your interpersonal communication.

For Individuals

- Consider your goal and audience. You want to tailor your communication to your goal and audience, both in terms of style (see different styles of communication, earlier in the chapter) and the medium (email, telephone call, face-to-face, etc.). So think about what it is you want to achieve (are you asking for something, for example?), what action you want the audience to take, and then decide on the best method to achieve that goal.

- Brush up your written communication skills. Spelling and grammar matters, so always proofread any written communications before you hit send. Make sure your message is clear, concise, and free of waffle. Also, be wary of sarcasm because it's really hard to convey it in writing. If you're wanting to portray a particular emotion, try using emojis to reinforce the underlying sentiment.

- Tell a story. When you're communicating a lot of information—such as in a presentation or report—try to tell a story, rather than relying on straight facts and figures. Before you start, distill the information down to one core message, then craft a narrative that really drives that message home. Also try to link the information to real-world events or anecdotes that, for example, bring a struggle to life or demonstrate *why* the information matters—basically, you're aiming to attach emotions to the information you're delivering. But keep it simple and authentic. You're not looking to win an Oscar here!

- Overcommunicate. One key thing I've learned from experience is to overcommunicate. Even if you think you've made something crystal clear, say it again. And maybe again after that. This leads to the next point.

- Recap information. At the end of a meeting, call, presentation, or even a written report, always quickly summarize the key points (much like I do at the end of each chapter in this book). Also take the opportunity to reiterate any actions that others need to take. Then check for understanding by asking whether it all makes sense and whether anyone needs further clarification.

- Detach from your phone. It's so important to practice active listening. This is why I make it a rule never to have my phone on in a meeting—it's just not fair to everyone else in the room. I highly recommend either leaving your phone elsewhere or switching it off for meetings.

- Show that you're listening. You can do this by nodding, making encouraging "mmm-hmm" sounds, making eye contact, taking notes, and asking follow-up questions.

For teams

Here are some additional pointers if you're working as part of a remote team:

- First things first, revisit the pointers from Chapter 9 on collaborating effectively in remote teams. Collaboration and communication are hugely intertwined.

- Acknowledge that communicating remotely is different from communicating in person. This is especially important if you're used to communicating with your colleagues face to face but now find yourself spending more time working remotely. You may need to tweak communications in line with the new working arrangements—for example, holding meetings at different times to accommodate home life, or making meetings shorter. An hour-long video meeting between 12 people, for instance, could get cumbersome, even if that's how long the meeting used to take in person.

- Use the right communication tool for the right task. For example, you might use Slack for informal conversations, email for formal work-related requests and info, and project management software for project status updates.

- If in doubt, pick up the phone. Especially when you're primarily conveying emotion rather than information, you may find it's better to jump on a call rather than send a written message.

- Use video chat as often as possible. In audio-only calls, you lose those precious nonverbal signals that are key to understanding. So encourage others in the team to embrace video as the default remote meeting format, and for one-to-one chats.

- Remember that you're on camera. When you're on a video call, always remember others can see you! So no reaching for your phone. Watch your body language and facial expressions. And make plenty of eye contact with the camera.

- Don't forget to make time for casual interactions, of the sort you'd have every day if you were in the office.

A quick note for organizations

The above tips for remote teams apply to organizations just as much as individuals, and will become especially important as more teams transition to remote working. Technology is your friend here, so invest in the technology platforms that will help your people communicate effectively from a distance—and perhaps set guidelines on which platform is most appropriate for which type of communication.

Do also revisit the practical steps in Chapter 9 (collaboration) as these will help you boost communication within the business.

Key Takeaways

In this chapter we've learned that:

- Interpersonal communication is simply the exchange of information, emotions, and meaning between people. This spans oral, written, and nonverbal communication, as well as listening. Nonverbal communication makes up the overwhelming majority of understanding.

- There are different styles of communication (assertive, aggressive, passive, passive-aggressive, and manipulative), so part of being a good communicator involves recognizing your own communication style, and that of others, and adapting your communication accordingly.

- Interpersonal communication is a skill that's prized by employers. In fact, demand for social skills like communication has been on the rise for decades. What's more, communication is one of those skills where humans massively outperform machines—talk about future-proofing your career!

- Simple ways to improve your communication include practicing active listening, overcommunicating to ensure understanding, and telling a narrative story instead of bombarding people with facts and figures. Remote workers must ensure they maintain lines of communication by, among other things, using video where possible (to ensure the nonverbal subtleties aren't lost) and making time for informal conversations.

I've mentioned remote working a lot over the last couple of chapters, because I believe it's one of the key factors that will shape the future of work. As well as working in remote teams, more and more of us will be working as gig or contract workers. This brings us neatly onto the next future skill.

Notes

1. Nonverbal Communication: How Body Language & Nonverbal Cues Are Key; Lifesize; https://www.lifesize.com/en/blog/speaking-without-words/

2. The growing importance of social skills in the labor market; Harvard; https://scholar.harvard.edu/files/ddeming/files/deming_socialskills_qje.pdf

CHAPTER 11
WORKING IN GIGS

In their book *The Human Cloud*, Matthew Mottola and Matthew Coatney describe a new way of work in which AI and the freelance economy combine to transform the world of work. They argue that traditional full-time employment will be a thing of the past, as organizations shift to hiring people on a contract basis, with those contractors mostly working remotely.

I tend to agree with their vision, which is why I'd advise everyone to prepare for a future in which more people work as "free agents" rather than traditional employees. And even if you happen to remain in traditional employment, the freelancer's ability to be flexible and take responsibility for their own professional development is something we can all learn from. Indeed, the need for everyone to take charge of their own skills is one of the main factors that drove me to write this book in the first place (and is probably why you picked it up). In other words, if you think the "gig economy," as it's so often called, doesn't apply to you, think again. This trend is relevant to us all.

But before we get into the details, we can't ignore the fact that the gig economy has something of a bad rep. The phrase conjures up images of people working subminimum-wage jobs with next to no job security, while huge, exploitative platform businesses profit from their labor. While that sort of work is arguably what catapulted the gig economy into the mainstream, the gig economy spans all kinds of independent, freelance, project, and contract work. In fact, you might be surprised by the most prevalent jobs that gig workers do (more on this coming up). Maybe "gig economy" will be replaced with a better name in future—perhaps something less opportunistic-sounding, such as "project-based economy" or simply "freelance economy"—but for now, know that I'm using the phrase "gig economy" to encompass all kinds of independent working.

What Is the Gig Economy?

You know what it is, of course. You probably participate in it in one way or another—as a freelancer, as an employer, or just as someone using platform services like Deliveroo or DoorDash. But let's agree on a formal definition anyway. The gig economy is where businesses and independent workers engage in short-term contracts or freelance work, as opposed to permanent employment. It started with digital platform businesses that connect freelancers with people who want their services—think platforms like Uber, TaskRabbit, and the like—but is now increasingly used to refer to any kind of freelance or short-term working arrangement.

Some people work independently because they love the freedom. For others, it's a side hustle that helps to boost their finances in addition to regular employment. And of course, some do it out of necessity, for example, after they've been made redundant.

What sort of gigs are we talking about? You may be surprised to learn that among the biggest segments of the freelance economy are creative and knowledge-intensive work. Globally, 59 percent of gig workers are engaged in design, technology, and IT jobs; following behind that are jobs in multimedia production, content writing, and marketing, which combined account for 24 percent of gig work.[1]

Overall, the gig economy numbers are pretty staggering. More than a third of US workers (around 57 million people) are already engaged in the gig economy, which is expanding much faster than the US economy as a whole.[2] And the number of US gig workers is set to grow to 86 million by 2027. That's more than half the working population of America.[3] In the UK, there are more than 4.3 million self-employed people (including business owners), with 1.9 million of those being full-time freelancers.[4]

Technology has of course played a pivotal role in the gig economy's rise. Uber, for example, wouldn't exist without the underlying platform to connect drivers with customers. Technology has also enabled remote work in general, thereby opening up a wide range of jobs to project-based, free-lance work, including the knowledge-intensive and creative work that makes up the majority of the gig economy. Economic factors have also played a role, not least the financial crisis of 2008, in which many people were thrust into unemployment or underemployment. This pattern repeated during the pandemic when, sadly, many learned that being a permanent employee isn't as secure as we like to think.

Bottom line, the gig economy is undergoing a huge growth spurt. And I'm not surprised. If you think about it, it's a natural progression of work. The days of a nine-to-five "job for life" are long gone. The reality today is that most of us will work for many companies over the course of our career, perhaps working flexible hours or from home, and often working with colleagues from all over the world. The gig economy is merely an extension of this, building on the idea that we as workers will move from job to job and team to team over the course of a long career—it's just that some or all of those jobs in future may not be on a permanent contract. (The very name "permanent contract" is fairly redundant considering many people only stay in a job for a few years at a time.)

Certainly, in my consultancy work with companies, I've noticed that organizational boundaries are becoming more porous. Many organizations are just as likely to engage independent workers as they are to hire permanent employees, and these independent workers slot into their teams and projects, collaborating with others as if they were full-time members.

Why It Matters: The Benefits of the Gig Economy

With so many people working in the gig economy, individuals and businesses alike need to get to grips with the changing nature of employment.

Work without boundaries

For freelancers, the gig economy offers some enticing benefits. The traditional employment model of working for one employer at a time, usually within commuting distance of where you live, is one of necessity. Until technology made remote work possible, working fairly close to where we live—and limiting our choice to local(ish) employers—was the norm because it was the only option. But with the growth of the gig economy, there are, in theory, no such boundaries. The next company you work for could be located at the opposite end of the country, or indeed in another country altogether. You no longer have to live in or near a certain city because "it's convenient for work."

Gig work is also more flexible than traditional employment. (You can read more about flexibility, and why it's a future skill, in Chapter 12.) Independent work fits around other factors like family life, creative passions—or just the hours when you're most productive, which may or may not be the traditional nine-to-five routine. According to data from Ohio University, the average person is productive for less than three hours out of an eight-hour workday, and the top 10 percent of highly productive employees take 15- to 20-minute breaks every 52 minutes.[5] It makes you wonder why the nine-to-five routine with a one-hour lunchbreak in the middle has reigned supreme for so long.

Independent workers also have the freedom to choose which projects they work on, which projects best align with their interests, and which projects will help them achieve their personal and professional goals. And depending on your industry, you may earn more as an independent worker, because you're free to set your own rates. Even better in these unpredictable times, that income comes from multiple sources, not just one source. So if you lose one client, you have the other income streams to fall back on, whereas in traditional employment if you lose your one job, you probably face a lot of financial uncertainty.

Of course, it's not all rosy. Depending on where you are in the world, independent workers rarely enjoy the same rights and protections as those on the payroll, such as minimum wage, paid holiday, sick leave, health insurance, and a pension. Hopefully we'll see greater recognition for gig workers in future (see the 2021 UK Supreme Court ruling that Uber drivers are entitled to minimum wage and paid holiday).[6] In fact, some experts are suggesting that the future of freelancing may involve "portable benefits"—benefits that are tied to individual workers, not the employer, where businesses contribute to independent workers' benefits based on how much or how often they work for them.

Overall, I believe the gig economy brings more benefits than downsides. I would say that—I'm a gig worker myself! But the majority of independent workers agree with me; in the US, for example, 79 percent of gig workers say they're happier being a free agent than being employed in a traditional job.[7]

Tapping into a global talent pool

It's not just freelancers who benefit. For employers, the gig economy means the talent pool is now truly global. Your freelance hires could be located anywhere in the world. You're no longer limited to those who live in the same city, or those who are willing to move. You're no longer at the whim of local labor markets. You know the oft-quoted saying that people are a company's most important asset? So why should businesses limit themselves to the talent in one geographic area? This opening up of the talent pool is also a positive thing for diversity and cultural intelligence (see Chapter 13).

Plus, there are very practical advantages around being able to adapt to market changes, by tapping into extra talent as and when you need it—in line with seasonal demand, for example, or economic fluctuations. This allows businesses of all shapes and sizes to remain competitive and continue to drive the business forward, in any market conditions.

How to Prepare Yourself for This Major Shift in Employment

Let's explore how individuals and employers can thrive in the gig economy.

For Individuals

Whether you're interested in becoming a freelancer or are one already and want to be better at it, there are plenty of steps you can take to improve your chances of success. (This is absolutely vital in a burgeoning gig economy, where you'll face increasing competition from around the world.) And even if you don't see yourself taking the leap into gig work anytime soon, do give this section a read, because even those in traditional employment can benefit from adopting a more entrepreneurial mindset and taking greater responsibility for their career development.

- The gig economy can be unpredictable, so you'll want to hone your adaptability. Turn to Chapter 12 for more on adaptability and flexibility, and see, Chapter 19 about embracing and managing change.

- Work on your interpersonal communication (Chapter 10), to build and maintain rapport with your clients and any other team members you work with. It's essential for winning repeat business. If you struggle with communicating in an assertive style, this is another thing you'll want to finesse—because, as a freelancer, it's up to you to confidently communicate what you expect and to be your own best advocate.

- Embrace the art of self-promotion, because doing the work is only one side of the freelance coin—the other side is *getting* the work.

Read more about this in Chapter 16, which is all about networking and creating your personal brand.

- Be willing to learn continually. In the gig economy, you're essentially selling your skills. So the more skills you have, and the more up to date they are, the better your chances of success. Read more about continual learning in Chapter 18.

- Build your own dream team around you. As an employee, you may have little or no control over who's on your team. But as a freelancer, you can build and strengthen relationships with whomever you want, but particularly those who complement your own skills, challenge you in a positive way, and genuinely help you succeed. These people could be fellow freelancers, mentors, clients—anyone who helps you be a better freelancer. This may include people you can outsource work to when you're overloaded.

- Remember that gig working is a business. You don't want to be working all hours only to find that your freelance work doesn't bring in enough cash to pay the bills. So keep a careful eye on your finances, specifically your profit margin, to ensure that gig work is financially viable for you. (You may want to brush up on your business skills with an online course.) It's also a good idea to diversify your freelance income stream as much as possible so that you're not dependent on one client to put food on the table.

- Document your work and progress, so you can both provide clients with regular updates and build a portfolio of success stories to share with potential clients in future. This can take many forms—a bullet journal, a notebook, a project spreadsheet—but the basic idea is to track how you spend your time and what you accomplish. This will help you determine what works and what doesn't, gauge how you spend the majority of your time, and give you a basic framework of notes that you can build out into detailed case studies in future.

- Find a routine that works for you. Are you, for example, at your most productive in the morning? Are you a night owl? Do you prefer short bursts of productivity, followed by a quick break, or are you someone who likes to really immerse themselves in a project for hours at a time? Basically, find out how and when you work best. It's key to staying motivated as a freelancer.

- Stick to a routine. Yes, in theory, we freelancers can work whenever and wherever we want. But the reality (for most of us at least) is that we thrive when we maintain some sort of regular routine. So when you discover what suits you, try to stick to it. This doesn't have to be nine-to-five—far from it. But some level of consistency will help you stay motivated. And that goes for your working environment, too. Having a set place of work, even if it's one end of the kitchen table or a corner of your favorite cafe, can help you quickly settle into "work mode."

- When it comes to finding gig work, networking is obviously key (again, see Chapter 16). But there are also numerous apps and platforms that connect freelancers with potential customers. You may or may not find these worthwhile, but if you do make use of them, be sure to keep your profile and portfolio up to date.

- Start small and build your confidence. If giving up a steady job and leaping into a freelance career fills you with terror, remember that gig work can be a great side hustle, something you do part time alongside regular employment. It can even help you gain skills in a new or adjacent field—potentially leading you to an interesting new career in future.

Clearly, success as a gig worker relies on so many related skills from this book, so as well as those chapters mentioned above, do revisit critical thinking (Chapter 5), emotional intelligence (Chapter 7), collaboration (Chapter 9), time management (Chapter 17), and looking after your physical and mental health (Chapter 20).

For organizations

It's clear that the workforces of the future will increasingly feature independent workers. And that means organizations must start planning for a more blended workforce. Start building your freelancer network now, in other words. Even if you don't regularly dip into a freelancer pool at this time, chances are you'll need to in future.

When you do find brilliant freelancers, do what you can to keep them. As more companies dip into the gig economy, you'd better believe there'll be stiffer competition for the best talent. So be sure to offer competitive pay and steady work to your best freelancers, pay them on time, and show appreciation for a job well done. You want freelancers to feel like a valued part of the team.

With this in mind, it's also a good idea to design an onboarding process for independent hires, a bit like the onboarding you'd do for full-time employees. This should include giving them access to the tools and information they need to do their job well, educating them on your company culture and values, and properly introducing them to the people they'll be working with. And just as with full-time employees who work remotely, people who manage freelancers must maintain those lines of communication and make time for informal chats.

Also, try to offer as much flexibility as you can to your freelancers. Recognize that they won't necessarily want to work during office hours or every day of the week. And while most freelancers prefer to work from home, a café, or a dedicated coworking space, those who are based locally may actually want to spend some time in the office with their non-freelance colleagues. You may therefore want to have some hot-desking spaces where freelancers can occasionally stop by for a day of office working.

Key Takeaways

In this chapter we've learned that:

- Organizational boundaries are becoming more porous, and businesses are just as likely to hire talent on a short-term basis as they are to invest in traditional hires. In the future, full-time, permanent employment will no longer be the norm.

- While the gig economy started with platform businesses like Uber and TaskRabbit, the phrase is now commonly used for all kinds of independent, freelance, project, and contract work. Creative and knowledge-intensive work makes up the majority of the gig economy.

- More than a third of US workers are already engaged in the gig economy, and this is only going to increase. As such, individuals and businesses need to prepare for a future in which more people work as free agents. We must all learn how to thrive in the gig economy.

- Even those who remain in traditional employment can learn from the gig economy trend—in particular, learning to be more entrepreneurial and to take a proactive approach to personal and professional development.

Flexibility is one of the biggest draws to the freelancer life, so let's dig a little deeper into that topic, and see why all of us—not just independent workers—need to embrace adaptability.

Notes

1. Gig Economy—The Economic Backbone of the Future?; Brodmin; https://brodmin.com/case-studies/gig-economy-case-study/

2. Will The Gig Economy Become The New Working-Class Norm; *Forbes*; https://www.forbes.com/sites/forbesbusinesscouncil/2021/08/12/will-the-gig-economy-become-the-new-working-class-norm/

3. Gig Economy—The Economic Backbone of the Future?; Brodmin; https://brodmin.com/case-studies/gig-economy-case-study/

4. The Self-Employed Landscape in 2020; IPSE; https://www.ipse.co.uk/policy/research/the-self-employed-landscape/the-self-employed-landscape-report-2020.html

5. Is an eight-hour workday productive?; SBAM; https://www.sbam.org/is-an-8-hour-work-day-productive/

6. UK Supreme Court's Uber Ruling May Prompt Gig Economy Changes; SHRM; https://www.shrm.org/resourcesandtools/hr-topics/global-hr/pages/uk-uber-ruling-changes.aspx

7. Gig Economy—The Economic Backbone of the Future?; Brodmin; https://brodmin.com/case-studies/gig-economy-case-study/

CHAPTER 12
ADAPTABILITY AND FLEXIBILITY

"Change is the only constant." Boy, did Greek philosopher Heraclitus hit the nail on the head. Considering he was alive around 500 BCE, it just goes to show that change is an enduring fact of life for humans. Yet despite its constant presence, many people struggle with change—or at the very least are wary of it. Change is uncomfortable. Change transports us from the known and safe into the unknown. It's no wonder many people don't exactly welcome change with open arms.

But in the workplaces of the future, change is going to be even more of a driving factor than it is today. New technologies and rising automation, the rapidly evolving pace of work, regular and vast business disruption, from extreme weather events to a global pandemic—these are all things that individuals and organizations have had to cope with in recent years. Think back to life 10 or even 5 years ago and it's clear that so much has changed in a short period of time. Looking ahead, I'm certain that the pace of change is only going to accelerate. And that means we must all develop the mental resilience to thrive amidst constant change. Adaptability is key to this.

In this chapter, I'll look at the mindset side of change—or what it takes, mentally, to embrace change. Turn to Chapter 19 to read about the more practical aspects of managing change in the workplace.

What Are Adaptability and Flexibility?

Adaptability is our ability to adjust to new conditions. If someone is described as "flexible," it essentially means they have a high level of adaptability. (And yes, adaptability can be measured, much like IQ and EQ. More on that coming up.) If someone doesn't respond to change so well, it's because they are low in adaptability.

I really like how Tony Alessandra and Michael O'Connor describe adaptability in their book *The Platinum Rule*. They say adaptability is made up of two components: flexibility and versatility. The first component, flexibility, is about mindset—it's your *attitude* to change. (This isn't to say flexible people are never skeptical or wary of change. But they are *willing* to change, which is key.) The second component, versatility, is about aptitude—as in, are you actually *able* to change? People with adaptability are both flexible and versatile; they're willing and able to change.

It's a fascinating way to approach adaptability, not least because it implies we have some control over how we react to change—being willing to change is a choice that we can all make, or not, and is something we can all work on, or not. Therefore, we can all become more adaptable, if we're willing to work at it.

What does an adaptable person look like?

Focusing on the flexible mindset side of things, what does flexibility look like in practice?

- Flexible people see opportunities, not obstacles. As part of this, flexible people are willing to try a variety of solutions, and aren't just wedded to "plan A."

- Flexible people are open-minded, not just about change itself but about all sorts of things: new ideas, other people's points of view, other people's experiences, different cultures and values, you name it. Flexible people are willing to listen to others, even if it means challenging their own opinions and preferences.

- Flexible people are curious, and willing to continually learn new things (see Chapter 18). I'd say this also includes critical thinking (Chapter 5), because critical thinkers are curious about the information that is presented to them, and always seek to find the full picture.

- Flexible people persevere and stick at it even when the going gets tough. In order to do this, flexible people are generally better able to cope with failure, manage stress, and stay positive.

- Flexible people may also be creative (Chapter 8), good decision makers (Chapter 6), and good collaborators (Chapter 9), because change often means you have to find novel solutions to problems and work with others to find new ways of doing things.

- Flexible people tend to have strong interpersonal skills (Chapter 10), because communicating well with others often involves adapting your communication style and behavior. I'd therefore say flexible people also rank highly for emotional intelligence and empathy (Chapter 7).

What about the inflexible people, how might we spot them in the workplace?

- Inflexible people are close-minded, and reject ideas that don't align with their own preferences or approach.

- Inflexible people would rather win than compromise (see the aggressive communicator, Chapter 10).

- Inflexible people may get defensive if they think change threatens their preferred way of doing things.

Introducing the "adaptability quotient," or AQ

Much like cognitive intelligence (IQ) and emotional intelligence (EQ), adaptability can be measured. And this is the idea behind the adaptability quotient, or AQ—literally a measurement of one's ability to respond to change. As with IQ and EQ, AQ is another key indicator of success. Indeed, in today's rapidly changing workplaces, some experts argue that AQ may be a more accurate indicator of performance. When you think of adaptability as simply another form of intelligence, just like IQ or EQ, it becomes startingly clear that, as with the other forms of intelligence, adaptability can be improved. I talk more about boosting your adaptability at the end of the chapter.

Do an internet search for adaptability quotient and you'll see there are various surveys you can take online to establish your AQ score. (And for organizations, there are tools that can help you assess AQ in your workforce.) In particular, I recommend AQai, which provides a robust tool for measuring, understanding, and improving your AQ; unlike standard surveys, it's conducted using a chatbot.

Why Do Adaptability and Flexibility Matter?

We'll undoubtedly see an acceleration of change, largely driven by technology and automation. Jobs will be transformed. New jobs will arise that we never could have imaged (hello, social media influencer). Others will disappear entirely (farewell, video rental store clerk). According to McKinsey, up to 375 million people may need to switch occupations and learn new skills by 2030.[1] That's a lot of change coming our way.

On top of this, there are wider global events to contend with, like the COVID-19 pandemic, war, or political instability. This isn't me being gloomy; I'm just saying that our world can be, and often is, unpredictable.

The personal benefits of becoming more adaptable

Adaptability gives us the mental resilience and the practical ability to respond to unpredictable situations and adjust to new conditions, not just at work—although adaptability is a key driver of workplace success—but also in everyday life.

We all have to weather challenges in life, from the big stuff like relationships ending and the death of loved ones to smaller obstacles like a holiday being canceled due to a new COVID variant or walking into a party and realizing you hardly know anyone. Being adaptable won't stop these sorts of things from happening, but it does make it easier to cope with the challenge, because (no doubt about it) change is stressful. It's hard work, isn't it, adapting to a new situation, trying to learn a new skill, or attempting to unlearn a bad habit? (Anyone who has ever given up smoking or biting their nails can relate.) Being mentally flexible helps to dull the anxiety that so often comes with new things. And as such, it's a really powerful skill to build.

There's also evidence that being adaptable leads to increased happiness and life satisfaction—not just because you can cope with the bad things that life throws at you, but also because you're better placed to find meaning and joy, regardless of your circumstances.[2] In other words, if you can embrace change, you're more likely to be happy and satisfied, whatever the situation.

Adaptability also makes you more desirable as an employee. It shows employers that you can cope with changing circumstances, that you're open to ideas and are willing to learn new things. All are valuable qualities for today and the future.

Why it matters to businesses

What about the benefits for organizations? Having a highly adaptable workforce and leadership is what allows a business to adapt to disruption, respond to new trends, cope with changing business models, and overcome challenges. To put it bluntly, adaptability separates the winners from the losers. You may think the saying "adapt or die" is laying it on a bit thick, but research indicates that many leadership failures are the result of an inability to adapt and let go of the "this is how we've always done things" mentality.[3] Think of those businesses that have failed to adapt to changing consumer expectations or new technologies—companies like Kodak, Blockbuster, and Blackberry. All of them failed to let go of the "old way" of doing things and embrace new business models, new partnerships, or new technologies that could have saved the business. Meanwhile, more agile organizations like Apple (which was perfectly happy to torpedo its own best-selling product, the iPod, by launching a smartphone that could also play music) are just plain better equipped to survive in our rapidly changing world.

What's more, offering flexibility in the workplace—working from home, flexible hours, and so on—will drastically improve your employer brand. Consider that more than 80 percent of UK employees feel flexible working makes a job more attractive. And flexibility is particularly attractive to millennials, of whom 92 percent say flexibility is their top priority when job hunting.[4]

How to Become More Adaptable and Flexible

Let's explore some practical ways that individuals and organizations can cultivate a more flexible mindset and, in turn, improve their adaptability.

For individuals

- Be open-minded and open to new ways of doing things. You could, for example, look at a situation from multiple different viewpoints. As part of this, listen actively to other people's opinions.

- Learn to "unlearn," by which I mean don't be afraid to let go of old information and old ways of doing things, to make space for new information and methods. In our rapidly changing world, what worked yesterday isn't necessarily going to work tomorrow or the day after that. According to the founders of AQai, this ability to actively unlearn can boost adaptability by a whopping 40 percent.[5] If you want to get better at unlearning, they recommend you explore opposing perspectives and patterns of behavior, even if they conflict with your existing knowledge.

- Keep up with the latest trends in your industry. This will help you stay open to new ideas, and uncover exciting new ways of doing things.

- Try implementing a small change at home, such as reorganizing the garage or changing the layout of your living room. It's a reminder of how invigorating change can be. At work, you could freshen things up by changing the running order of a regular meeting, meeting in a new space, or even just reorganizing your desk.

- Ask plenty of "what if" questions. You can use this technique to think about what *might* happen before it actually happens. For example, "What if X aspect of my job changes to Y?" How would you positively navigate the change?

- Step outside of your comfort zone and put yourself in situations that positively challenge you. If you'd normally avoid a certain social situation, for example, give it a go. If you usually shy away from taking on additional responsibilities and projects at work, actively put yourself forward. Same goes for training opportunities. Say yes to more stuff, basically. Even small steps can really add up over time.

- Accept that trying new things may lead to failure, and that's okay. It's all part of learning. As the saying goes, try and fail, but don't fail to try. If you've tried and failed at something in the past—kicking a certain habit, for example—why not revisit it again, this time trying a new technique?

- Notice the small changes going on around you. If you struggle with adaptability, it can be tempting to ignore change. So pay attention to the small changes that take place around you all the time—from your colleague's new haircut to an updated company policy. That way, when bigger change comes along, you'll be better equipped to see it coming and you won't be caught off-guard.

- Don't freak out when things change suddenly. Yes, it's normal to find sudden change stressful, but you're in charge of how you respond to that stress. The best thing you can do here is take care of your physical and mental health. Head to Chapter 20 for more on this topic.

- Practice optimism. Things often don't go according to plan, but rather than focus on the things that haven't worked out, try to focus on the positive aspects. Adaptable people are generally more optimistic, and have a wonderful ability to find satisfaction even in not-so-positive circumstances. It's one of the reasons why adaptable people are happier. Even small steps like changing your language—for example, calling an "obstacle" an "opportunity"—can help.

- Learn from others who routinely demonstrate adaptability in their work and home lives. Note what strategies they deploy and what sort of language they use. You could even ask them for tips on how they cope with change.

- Finally, believe. Believe in your ability to roll with the punches and succeed in life, no matter what comes your way.

Don't forget to head over to Chapter 19 for practical strategies that will help you manage change.

For organizations

Business leaders, take heart. There are many practical things you can do to build a culture of adaptability and flexibility in your organization. One of the best starting points is to begin a conversation around flexible working arrangements. This can be as small as allowing regular remote working, or as big as switching to a four-day workweek. You can also reconfigure the office to create a more flexible workspace—think breakout areas where teams can conduct informal meetings and work collaboratively, or even hot-desk areas with standing desks. All of this can inspire people to adopt a more flexible mindset at work.

For me, an important part of flexibility is giving teams the freedom to achieve objectives in their own way. But of course this requires leaders to be crystal clear on the organization's objectives and values; only then can teams and individuals identify their own best way to contribute. This fosters creative thinking and inspires people to let go of "old ways" and try new things—but, crucially, this must be without fear of failure. In other words, your culture needs to be one that sees failure as a learning opportunity, not something to be punished.

No doubt you'll come across people and teams that are resistant to change. I talk more about managing change in Chapter 19, but from a mindset point of view, know that people are generally more open to change when they understand *why* that change is necessary. As with any business change, always sell the benefits of the change: how it will make people's jobs easier, better, and so on.

When people show inflexibility, listen to their concerns and practice empathy, while again being clear on why the change is necessary. It can also help to clarify which elements of a job are more subject to change than others; for example, software is something that changes regularly, while health and safety protocols tend to remain more static (pandemics aside, of course!). Sometimes, just being forewarned that something is subject to change helps people respond more positively when the time comes.

Finally, do try to celebrate adaptability and give positive reinforcement when individuals and teams demonstrate high adaptability.

Key Takeaways

To briefly recap the key points on adaptability:

- We're going to see an acceleration of change, largely driven by technology and automation. Jobs will be transformed, new jobs will arise, and others will disappear entirely. As many as 375 million people may need to switch occupations and learn new skills by 2030.

- With so much change coming our way, we must all develop the mental resilience to thrive amidst constant change, not just at work—although adaptability is a key driver of workplace success— but also in everyday life.

- Adaptability is made up of two components: flexibility and versatility. Flexibility is about mindset (your *attitude* to change) while versatility is about aptitude (your *ability* to change). People with high levels of adaptability are both flexible and versatile; they're willing and able to change.

- Your adaptability quotient (AQ), can be measured, just like IQ and EQ. It can also be improved. In fact, there are many steps you can take to cultivate a more flexible mindset and, in turn, improve your adaptability. You can keep up with the trends and changes in your industry, push yourself outside of your comfort zone (even in small ways), and practice optimism—to name just a few examples.

Among the many changes in the workplace, we're seeing greater emphasis on diversity, equity, and inclusion. Which brings us to another key future skill: cultural intelligence and diversity.

Notes

1. Jobs lost, jobs gained: What the future of work will mean for jobs, skills and wages; McKinsey; https://www.mckinsey.com/featured-insights/future-of-work/jobs-lost-jobs-gained-what-the-future-of-work-will-mean-for-jobs-skills-and-wages

2. The One Trait You May Not Realize Will Make You Happier; HuffPost; https://www.huffpost.com/entry/adaptable-people-habits_n_5508678

3. The importance of adaptability skills in the workplace; Training Journal; https://www.trainingjournal.com/articles/features/importance-adaptability-skills-workplace

4. Increase in demand for flexible working; Capability Jane; https://capability jane.com/about-us/flexible-working/

5. How To Thrive During Changing Time—The Adaptability Quotient You Need To Know And Improve; *Forbes*; https://www.forbes.com/sites/victoriacollins/2020/09/15/how-to-thrive-during-changing-time--the-adaptability-quotient-you-need-to-know-and-improve/

CHAPTER 13
CULTURAL INTELLIGENCE AND DIVERSITY CONSCIOUSNESS

We hear a lot about authenticity these days, largely in the context of being an authentic brand or an authentic leader. But it's just as important at an individual level, too. We should all be free to be our authentic selves at work, to bring our "whole selves" to work, so to speak, without fear of discrimination. For this to happen, however, we need workplaces that recognize and celebrate our many differences, cultural and otherwise. We need workplaces that truly reflect the diversity of our societies.

I had an early lesson in cultural diversity when, as a young man, I left my home in Germany and spent a year in England improving my grasp of English. On the surface, life in Germany and life in Britain aren't so different, but I was struck by so many contrasts. The food and drink. The humor. The British obsession with the weather. The fact that people patiently stand in line (and, jumping the line is not an option)! Later, I spent some time in Hong Kong, which, as you can imagine, was an even greater culture shock. Both experiences taught me that there are so many beautiful and fascinating ways in which our societies differ.

These days, I'm seeing organizations become much more representative of such differences. This is why I consider cultural intelligence and diversity consciousness to be a vital future skill.

What Are Cultural Intelligence and Diversity Consciousness?

Diversity simply refers to the many ways in which people can differ, including race, gender, culture, age, religion, sexual orientation, political beliefs, socioeconomic status, (dis)ability, and so on.

When we encounter things that make us different, conflicts and misunderstandings can easily arise. Eye contact is a great example. In my culture, making eye contact is a sign of respect and shows that I'm paying attention. But in other cultures, it can be a sign of aggression or even sexual interest, so to make sustained eye contact with someone while they're talking is considered rude or inappropriate. There's no point pretending that we're all the same or that differences don't matter. They matter a great deal, not least because it's the differences between us that spawn prejudice and discrimination. This means that part of creating a fairer world, and fairer organizations, entails being respectful of our differences, as opposed to glossing over them.

Awareness is the first step

This, in a nutshell, is what we mean by diversity consciousness: a basic awareness of diversity, a recognition that workplaces (and societies) are becoming more diverse, and that this diversity is a good thing.

If diversity consciousness is about *awareness*, cultural intelligence (also known as cultural quotient, or CQ for short) refers to our *ability* to relate to others from different backgrounds and work effectively in diverse situations. It's the diversity equivalent of cognitive intelligence (IQ) and emotional intelligence (EQ). And just like IQ and EQ, cultural intelligence can be an important predictor of success in the workplace. It can also be assessed and improved, but I'll get to that later in the chapter.

Cultural intelligence therefore goes a step beyond diversity consciousness, because someone who is culturally intelligent is not just aware of diversity—they're able to adapt and relate to people from all sorts of backgrounds. This encompasses several skills and capabilities, including:

- Being curious and open-minded, and using critical thinking (Chapter 5)

- Being empathetic and emotionally intelligent (Chapter 7)

- Being motivated to work well with others (Chapter 9)

- Being able to communicate effectively (Chapter 10)

- Being able to adapt behavior as and when needed (Chapter 12)

In this way, many of the other future skills in this book contribute to cultural intelligence, and vice versa.

Diversity, equity, and inclusion

Focusing on "diversity" alone can be misleading, which is why you'll commonly (or at least hopefully) see organizations talking more about diversity, equity, and inclusion (DEI) these days. If diversity refers to the many ways in which people can differ, equity is about providing fair access, treatment, and opportunity for people from all backgrounds. And inclusion is the extent to which people feel they belong and are valued— that they feel welcomed, that they have a voice, and that they're included in decision-making. In other words, you may have a diverse organization, but that doesn't mean it's an equitable or inclusive one—not if the opportunities for advancement aren't available to all, or if people don't feel valued or respected.

I could write a whole book on the subject of DEI and how organizations should tackle it. (Actually, it's better left to others who are much more qualified!) But I wanted to flag that, while this chapter is about *diversity consciousness* and *cultural intelligence*, we also need business leaders to ensure that organizations are equitable and inclusive. So when I refer to "diverse" workplaces in this chapter—particularly when it comes to the benefits of diversity—know that I'm talking about organizations that are also equitable and inclusive.

Why Do Cultural Intelligence and Diversity Consciousness Matter?

Our world is more connected than ever, which means that, every day, many of us encounter people who have different backgrounds and different life experiences than our own. I work with brands from all around the world, for example. But even if your business doesn't cross geographical boundaries, your colleagues and clients will almost certainly represent a diverse range of cultures, ages, ethnicities, economic statuses, and so on.

Collaboration across boundaries—whether geographical, cultural, political, or whatever—is essential if you want to succeed in the 21st-century workplace. It's essential if you want to be an effective team player, a strong communicator, a good leader, an emotionally intelligent colleague, and many of the other future skills that feature in this book. It may even affect your likelihood of securing a job (more on this coming up).

But what about the benefits to businesses? Of course there's a moral imperative to ensure organizations better reflect our increasingly diverse societies. And, let's be honest, diversity looks good for organizations. (Put it this way: can you see any business shouting about their *lack* of diversity?) But the benefits of diversity go way beyond the moral and marketing.

Diversity is good for business performance. There's a wealth of research to support this, such as a Boston Consulting Group study that found a significant correlation between diversity and innovation (companies with diverse management teams reported innovation revenue that was 19 percentage points higher than companies with below-average diversity).[1] Or there's the evidence that organizations with more women in corporate leadership positions are more profitable (a 30 percent female share of the C-suite translates into a 15 percent increase in profitability for the average firm).[2] What's more, this relationship between diversity and the likelihood of financial outperformance is getting stronger over time—and the greater the representation, the higher the likelihood of outperformance.[3]

Bottom line, organizations with diverse workforces and leadership are more likely to be successful. That's largely because diverse teams enjoy a broader range of perspectives and ideas than homogenous teams—and again, I'm not just talking about ethnic, cultural, and gender diversity, but also factors like age, political beliefs, and a myriad of other differences. Such diversity of talent allows businesses to be more innovative and adapt to increasingly diverse marketplaces—for example, a diverse workforce is better equipped to develop more inclusive products and services, and to better serve customers.

You get it—diversity is an asset. But it also presents challenges. As an example, according to one study, 90 percent of executives from 68 countries cited "cross-cultural" management as their biggest challenge in doing business across borders.[4] This is precisely why organizations need people who are culturally intelligent and conscious of diversity.

Some employers are even considering assessing candidates for cultural intelligence as part of the recruitment process. The University of Oxford, for example, has proposed that candidates will have to demonstrate their commitment to diversity, equity, and inclusion (for example, having called out a previous employer on diversity issues).[5] It's just a proposal at the time of writing—a proposal that has attracted its fair share of criticism, including being dubbed a "woke score"—but it certainly shows that cultural intelligence is fast becoming a must-have skill.

And it's not just employers who will be assessing candidates based on their diversity consciousness; candidates increasingly want to work for diverse employers. This is especially true among younger members of the workforce (79 percent of recent graduates rank a diverse workforce as a "very important" feature of potential employers).[6]

How to Boost Your Cultural Intelligence

Having read this far in the chapter, you'll know that workplaces are becoming more diverse, and why this is a positive thing. That's already a big step towards increased awareness. But what about the more practical side of things—the ability to work with people from a variety of backgrounds, cultures, and so on? The good news is that anyone can become more culturally intelligent.

Let's start with some tips for individuals, before looking at how organizations can boost cultural intelligence in the workforce.

For individuals

- Remember, no culture is superior to another. Underneath our cultural beliefs, religious beliefs, political beliefs, or whatever, people largely want the same thing—to live a decent life, do a good job, be happy, have a family (in all the forms that may take), have a safe roof over our head, achieve some level of financial security, cultivate friendships and relationships, and so on.

- Practice self-awareness and think about the sorts of biases that may narrow your vision or shape your behavior (see Chapter 5). Ask yourself how your own cultural background may influence your world view. Think also about the biases that may exist in your organization.

- Be curious about other people's perspectives and have interesting conversations with people who have different experiences and beliefs than your own. (Read more about this in Chapter 5.)

- Practice active listening. As with so many of the skills in this book, becoming a better listener can help you understand where people are coming from and deepen your knowledge.

- Consume content from around the world. For example, I read news stories and watch news broadcasts from countries like India, Russia, and China. It helps me understand how other cultures view the world. You can also watch movies or read books from other countries to widen your worldview.

- Watch a TV show or read books or articles that demonstrate opposing beliefs or viewpoints. For example, if your politics leans towards the liberal, you might tune into Fox News—it probably won't sway you from your political beliefs (in fact, it may cement them even further), but it'll certainly help you understand what matters to those at the other end of the spectrum.

- Go to a religious service for a denomination that's different from your own.

- Where possible, immerse yourself in different cultures and perspectives. If you travel to a new country, for example, get out there and wander through the food markets, ride public transport, and generally soak up the culture (it's more effective than purely reading about it). If you're lucky enough to be invited to a local's home, all the better.

- Work on developing other related skills from this book, including empathy, adaptability, and collaboration.

For organizations

- First, ensure your organization represents a diversity of talent.

- If you haven't already, make a leadership commitment to diversity consciousness and cultural intelligence, so that the importance of these skills filters down through all layers of the organization.

- Leaders should also reflect honestly on their own cultural skills—all leaders, but especially those who manage people across geographical boundaries.

- Consider doing an organizational CQ audit to assess cultural intelligence. There are numerous tools to help you do this, using questions like, "What cultures are represented in the business and which are unrepresented?" Or "Are people equipped to engage with diversity?" Auditing CQ will also help you assess progress over time.

- Create a strategy for boosting cultural intelligence and diversity awareness across the organization. This may include, for example, diversity training. (Clearly, this should all be part of wider DEI strategy that includes actions like reporting on diversity stats.)

- Factor cultural intelligence and diversity consciousness into the recruitment process, for instance, by asking candidates to describe an example of cultural intelligence from their previous employment.

- Factor cultural intelligence and diversity consciousness into the performance review process.

- Celebrate and reward the culturally intelligent behavior you want to see more of.

Key Takeaways

To summarize the key points on diversity consciousness and cultural intelligence:

- Diversity simply refers to the many ways in which people can differ, including race, gender, culture, age, religion, political beliefs, (dis) ability, and so on.

- These days, I'm seeing organizations becoming much more representative of such differences, which means we all need to be more conscious of diversity and work to improve our cultural intelligence.

- Diversity consciousness is the recognition that workplaces (and societies) are becoming more diverse, and that this diversity is a good thing. Cultural intelligence (also known as cultural quotient, or CQ for short) refers to our ability to relate to others from different backgrounds and work effectively in diverse situations.

- These skills matter because our world is more connected than ever and all of us are increasingly required to collaborate across boundaries (whether cultural, political, economic, or whatever).

Furthermore, diversity is good for business, translating into increased innovation and better financial performance.

- There are many ways to boost your cultural intelligence, including being more self-aware, immersing yourself in different cultures, consuming content from around the world, and more.

Treating people with respect and dignity, and being able to work well with people from all backgrounds is, for me, a fundamental part of living an ethical life, and running a successful business, for that matter. So let's dwell on the subject of ethics a little more, and learn why ethics is fast rising up the corporate agenda.

Notes

1. How Diverse Leadership Teams Boost Innovation; Boston Consulting Group; https://www.bcg.com/publications/2018/how-diverse-leadership-teams-boost-innovation

2. Study: Firms with More Women in the C-Suite Are More Profitable; Harvard Business Review; https://hbr.org/2016/02/study-firms-with-more-women-in-the-c-suite-are-more-profitable

3. Diversity wins: How inclusion matters; McKinsey & Company; https://www.mckinsey.com/featured-insights/diversity-and-inclusion/diversity-wins-how-inclusion-matters

4. Cultural Intelligence: The Essential Intelligence for the 21st Century; SHRM; https://www.shrm.org/hr-today/trends-and-forecasting/special-reports-and-expert-views/Documents/Cultural-Intelligence.pdf

5. Oxford University blasted for considering hiring based on 'woke score' of academics: report; Fox News; https://www.foxnews.com/politics/oxford-university-woke-score-academics-report

6. For younger job seekers, diversity and inclusion in the workplace aren't a preference. They're a requirement; *Washington Post*; https://www.washingtonpost.com/business/2021/02/18/millennial-genz-workplace-diversity-equity-inclusion/

CHAPTER 14
ETHICAL AWARENESS

You probably have a good grasp of what it means to be ethical. Most of us live life according to our own set of principles and values. And even if we can't eloquently put our own moral "code" into words, we certainly know unethical behavior when we see it.

So why highlight ethical awareness as a future skill? It's partly because of the changing nature of work. The digital transformation and wave of fourth industrial revolution technologies have given rise to a whole new set of ethical challenges to overcome—think of the dilemmas surrounding gene editing, for example, or artificial intelligence, or the use of people's personal data. Then there's the huge potential fallout from ethical missteps in today's world, where scandals can spread across the internet faster than you can say TikTok.

As we'll see in this chapter, businesses are rapidly cottoning on to ethics as a critical issue, which means they'll increasingly want to hire the kinds of people who can help them address ethical challenges.

What Do We Mean by Ethical Awareness?

Ethics can be described as your moral principles or values. To be ethical is to be conscientious about your choices—at work, and in everyday life—and to act with good intentions in mind. In this way, ethics is concerned with questions such as:

- What is right and wrong here?

- How can I live a good life?

- If I do this, will it harm others?

- How can our organization succeed without harming individuals, society, and our planet?

- How can our business actively improve the lives of individuals and communities, and make the world a better place?

We all have our own version of "ethical"

Ethics provides us with a moral roadmap of sorts. It gives us the tools to think about moral issues and guide our decision-making. And this can span anything from how we feel about abortion and capital punishment to the food that we eat and the products that we buy.

Speaking of issues like abortion and capital punishment, it's clear that ethics doesn't necessarily provide us with clear-cut answers; people often have different views on what's morally "right," and some issues are just plain messy or ambiguous. Ethics, then, is a sort of framework that helps us to decide for ourselves what's right and what's wrong.

The fact that one person's idea of ethical behavior differs from another's hasn't stopped philosophers coming up with ethical theories to tell right from wrong. For example, there's "virtue ethics," which states that living an ethical life means demonstrating virtues such as compassion and courage (virtue ethics considers traits like jealousy and selfishness to be unethical). Or there's the "utilitarianism" theory of ethics, which says that to be ethical is to maximize the amount of happiness around you and minimize the suffering.

Although the specifics may vary from person to person and theory to theory, ethics is ultimately about considering the interests of others, as opposed to serving our own desires to the detriment of others. To think ethically, then, is to think about more than yourself. It's to think about the common good.

Business ethics

If ethics is a set of principles that guide our decision-making and behavior, it's pretty clear that this is just as important in a business context as it is in everyday life. An ethical business is one that tells the truth, keeps its promises to customers and employees, treats people with respect, takes responsibility for its actions, and, ideally, aims to contribute something to the common good. An unethical business is one that puts profit above all other concerns, to the detriment of its employees, customers, society, and our world.

Ethical awareness in a business context therefore starts with thinking about the effect of business decisions on customers, employees, and other stakeholders (and yes, I'd include the environment as a stakeholder). Looking a little wider than that, ethical awareness also encompasses the organization's overall values, beliefs, and culture. Looking even wider, business ethics often means regulatory compliance—some laws, such as the Sarbanes-Oxley Act, are designed to ensure companies behave in a certain way (in this case, to prevent fraud). Other legal examples include environmental standards, minimum wage laws, and health and safety laws.

Bringing all this together, perhaps the best way to define business ethics is as a system that guides the values, beliefs, decisions, and behaviors of an organization and the people who work in it. For this to happen, ethics needs to be embedded into the very fabric of the organization, starting with ethical leaders who set the tone for how the business should behave. This then filters down so that everyone in the company has a clear idea on what's right and wrong (in terms of decision-making and conduct).

Technology and ethics

New technologies bring with them ethical concerns that can be tricky to address. AI is one of those technologies that presents significant ethical challenges, such as data bias and data privacy (see Chapter 2)—not to mention the morality of asking machines to make important decisions (which in the case of something like healthcare could mean life-and-death decisions). Gene editing is another technology that can be fraught with ethical dilemmas, such as who decides which genetic traits are "normal" and which are abnormal? Or could gene editing ultimately make society less accepting of people who are different?

Industries need people with ethical awareness to help them navigate these murky waters and ensure that technology is used for the benefit of all. This is why more and more businesses are hiring ethicists, and I'm not just talking about tech companies. The US Army, for example, has a chief AI ethics officer (as part of the army's artificial intelligence task force) who advises the army on incorporating ethics into AI design. If you raised an eyebrow at that, consider the potential implications of AI systems making decisions on, for example, target sites for drone attacks. To what extent should machines be involved in such decisions and where do we draw the line between AI-led efficiency and our moral responsibility as human beings? Companies of all types will be grappling with similar sorts of questions, such as "How can we get the best out of AI, while ensuring the well-being of our employees, customers, and other stakeholders?" and "Does this usage violate people's individual right to privacy?" and "Does this provide genuine value for the people we serve?"

All things considered, although ethical awareness is nothing new, I expect it to rise in importance in the coming years as society grapples with the potential uses and impacts of new technologies.

Why Ethical Awareness Is as Relevant as Ever

Aside from the rise of new technologies, there are many reasons why ethical awareness matters now more than ever, not least because of the huge potential backlash when businesses get it wrong.

There's a rich history of businesses making ethical missteps. Think of any big corporate scandal and chances are it stems from some sort of ethical violation, whether it's financial misconduct, child labor, mistreatment of employees (or even customers), discrimination, lack of transparency, or whatever.

Remember that video of a bloodied passenger being dragged off an over-booked United Airlines flight after he declined to be bumped from the flight that he had booked and paid for? Yep, that's exactly the sort of thing that sticks in people's minds. While the airline was well within its rights to ask a passenger to leave an overbooked flight, the brutal optics of it all, not to mention the company's subsequent apology (which lacked any real sense of remorse), created an overall impression of a company that doesn't think too highly of its customers. Such reputational damage can be enormously harmful to the bottom line. In the immediate after-math of that United video, $250 million was wiped off the company's market value.[1]

Or there's the Equifax breach, in which the personal data of 148 million customers was stolen by hackers, yet the company failed to report the breach for two months—a huge scandal that ultimately resulted in the company agreeing to pay up to $700 million in compensation.[2] Or there's British online fast fashion retailer Boohoo, which was embroiled in a scandal about working conditions in its factories—after which the company's share price almost halved.[3]

Why do ethical scandals have such a big financial impact? I think it's because we as humans are generally turned off by hypocrisy. Businesses talk all the time about "living their values" and being authentic. So when a business *says* they act a certain way and then we clearly see them *doing* something else, it leaves us with a bad taste in our mouths. As a result, we're more likely to take our money elsewhere. Indeed, 43 percent of consumers have stopped buying from a business because of unethical practices.[4]

Ethics also matters from an employer brand perspective because—no real surprise here—people want to work for ethical companies. Consider this: 73 percent of people say they wouldn't work for a company unless its values aligned with their own, and 82 percent would rather take a pay cut to work at an ethical company than get paid more to work somewhere with iffy ethics.[5] In other words, an ethical business is much more likely to attract the very best talent.

All this shows why ethics matters to businesses, which means it matters to us all as individuals who want to build successful careers. With ethics being high on the corporate agenda, it makes sense that businesses will want to hire people who have a strong ethical awareness, and who can act with ethical values in mind. Therefore, brushing up on your ethical awareness is a smart career move.

What's more, being ethically aware helps you live your own values, and stay true to what's important to you. It can help you weigh up difficult situations, at work and in everyday life, and make decisions that are right for you (circle back to Chapter 6 for more on decision-making).

How to Become More Ethically Aware

Let's start with some tips for individuals, before moving on to how businesses can boost ethical awareness.

For individuals

- First, identify and understand your own values. What is important to you? What personal qualities do you think are important in a person? How do you wish to behave, and for others to see you? Everything stems from this understanding.

- Then ask yourself, do you really follow your own ethics?

- Learn about ethics. There's no one single blueprint on how to live ethically, which, for me, only makes the topic more fascinating. So read up on ethics and delve into the various different schools of thought. It may help you define your own ethical code.

- Practice empathy (see Chapter 7). The ability to put yourself in someone else's shoes can really help you make ethical choices.

- Look for ways to help others in everyday life. Yes, you can do things like volunteering and giving to charity, but don't overlook the opportunities for small good deeds, like giving up your seat on a crowded train.

- Respect the rights, values, and beliefs of others. Forcing people to agree with you or to go along with what you want isn't ethical—even if you believe you're acting with their interests in mind.

- Avoid the temptation to impose your ethics on others. Remember, being ethical means different things to different people.

- It sounds obvious, but when you make a promise, keep it.

- When weighing up potential employers, ensure the company's values align with your own. (Think critically here; see Chapter 5. Try to see things as they *really* are, rather than how you want them to be.)

- Familiarize yourself with your current employer's code of ethics, so that you can act on these principles—and call out instances of unethical behavior. If you believe your employer is acting unethically, raise it with your manager in the first instance.

For organizations and leaders

- If you don't yet have a code of ethics to guide how people work together and engage with customers, make that your top ethics priority. Codes vary from business to business, but good guiding principles involve honesty, transparency, treating people with respect, and acting responsibly. Basically, think about the kind of behaviors you want to promote, rather than listing what you *don't* want to see.

- Build your ethical code into your company's induction program, so that new hires are clear on the company's values.

- Ensure all leaders model ethical behavior. Because, as the saying goes, "values are caught, not taught." This means leaders must act the way they want others to behave, and display qualities like honesty and transparency on a daily basis. Employees are much more likely to act ethically when they see leaders embodying the organization's values.

- Build ethical thinking into everyday decision-making by considering the impact of decisions on employees, customers, and other stakeholders.

- Give people the means to report unethical behavior in confidence, and to highlight ideas for doing things better. As part of this, you may need to train certain people on how to deal with and investigate ethics complaints.

- Hold people accountable for ethics violations. When people cross ethical boundaries, they must be held accountable. In the case of something like sexual harassment, you'll want to have a zero-tolerance policy, but some issues will be less clear-cut.

- Reward ethical behavior, for example, by praising someone in a team meeting for treating others with respect, or giving a formal award to those who act with integrity.

- Consider the ethical impacts of new technologies, particularly when it involves artificial intelligence and people's personal data. Every company should build ethics into their data and AI strategies.

- And when things go wrong (as they do), let ethics guide your response to any crisis. This means being transparent about what has gone wrong, apologizing, making it right for the people impacted by the situation, and putting in place practices to ensure the same thing doesn't happen again.

Key Takeaways

In this chapter, we've learned that:

- Ethics is a set of moral principles that guide our decision-making and behavior. But being ethical can mean different things to different people (plus, some issues are just plain messy or ambiguous—meaning there isn't always a definitive right answer). Ethics is therefore a sort of framework that helps us to decide for ourselves what's right and wrong.

- Ethical awareness is as relevant today as ever—arguably more so given the ethical challenges that new technologies present. Plus, there can be huge blowback when companies make ethical transgressions—impacting not just the market value of the company, but also customer loyalty and employer brand.

- For individuals, ethical awareness matters both because employers will increasingly be looking for people with this vital skill—indeed, I'm seeing more and more companies hire ethicists—and also because it will help you identify your own values and stay true to what's important to you, both inside and outside of work.

- Practical ways to boost your ethical awareness include identifying your own moral principles, assessing whether you actually follow those principles in reality, and practicing empathy, honesty, and selflessness. What's more, when assessing potential job opportunities, look for companies that have similar values to your own.

I've already mentioned how ethical businesses rely on ethical leaders. Integrity is a vital skill in any leader—without integrity, it's very hard for people to respect their leaders (that goes for society at large, not just organizational leaders). Turn to the next chapter and let's drill into some other qualities that make for a great leader.

Notes

1. United loses $250 million of its market value; CNN; https://money.cnn .com/2017/04/11/investing/united-airlines-stock-passenger-flight-video/

2. Equifax agrees to a settlement of up to $700 million over 2017 data breach; The Verge; https://www.theverge.com/2019/7/22/20703497/equifax-ftc-fine-settlement-2017-data-breach-compensation-fund

3. Boohoo is ready to party but investors might not be as keen; *Guardian*; https://www.theguardian.com/business/2021/may/02/boohoo-ready-to-party-investors-may-not-be-as-keen

4. If ethics sells more, why are so many brands behind the curve?; Chris Arnold; https://www.linkedin.com/pulse/ethics-sells-more-why-so-many-brands-behind-curve-dr-chris-arnold/

5. Ethical Dilemmas: How Scandals Damage Companies; Western Governors University; https://www.wgu.edu/blog/ethical-dilemmas-how-scandals-damage-companies1909.html#close

CHAPTER 15
LEADERSHIP SKILLS

Think leadership applies only to those at the top of the ladder? Think again. The combination of factors that will shape 21st-century work—distributed teams, increasing diversity, humans transitioning to more creative tasks, the gig economy, fluid organizational structures, and so on—mean that leadership skills will be important not only for those in traditional leadership roles, but increasingly for those individuals throughout the company who are expected to lead, whether they're leading a project or an entire department.

What It Means to Be a Leader Today

As Jack Welch, former CEO of General Electric, wrote in his book *Winning,* "Before you are a leader, success is all about growing yourself. When you become a leader, success is all about growing others." This is the epitome of what it means to be a leader. Whether you're a CEO, an executive, a line manager, or a project lead, good leadership is about making sure other people can thrive. It's not about personal power; it's about serving the interests of others. And by doing that, you enable individuals and teams to deliver the organization's common goals.

Author and inspirational speaker Simon Sinek describes it as like being a parent, which really resonated with me. If you think about it, he's right. Being a leader means you're entrusted with the care and well-being of others—it's your job to help them grow, so that they can be the best they can be, even long after you've gone.

It's a simple definition of leadership—someone who grows others—but of course this requires lots of different skills, many of which we've explored elsewhere in this book (communication, collaboration, creativity, decision-making, flexibility, emotional intelligence, cultural intelligence, ethics, and so on). Leadership is thus a collection of skills, which makes this chapter a little different from the other chapters, which have generally focused on one skill at a time. In this chapter, we'll explore multiple qualities that contribute to effective leadership, focusing on those that aren't covered elsewhere in the book.

But before we move onto specific skills, let's briefly talk about the notion of a "natural-born leader." Is it true that leadership qualities are innate? In part, yes, some of the things we associate with great leaders—particularly things like sense of humor and charisma—are qualities that we're born with. But studies have shown that these innate traits account for only a small part of leadership; for example, one study conducted with twins found that only 30 percent of the variance in leadership role occupancy could be associated with innate qualities.[1] In other words, the majority of leadership qualities can be learned and improved—which means anyone can be a great leader. And in the 21st-century workplace, many more people, at all levels of the business, will need to demonstrate leadership abilities.

This is an important point to dwell on, because people often think that "leadership" refers to the CEO and executive team. But now, with the rise of the gig economy, and with many companies adopting flatter, more flexible organizational structures, leadership applies to more people than ever before. You may be overseeing a project that requires you to coordinate several team members. Or you may be a gig worker collaborating with other gig workers. Or you may be occupying a traditional management role. Whatever your job title says, this precious ability to bring out the best in others and help them thrive is vital amidst the changing nature of work. Therefore, everyone should be looking to cultivate leadership skills.

As Simon Sinek says in his book *Leaders Eat Last: Why Some Teams Pull Together and Others Don't*, "If your actions inspire others to dream more, learn more, do more, and become more, you are a leader." So forget the job title; a leader is anyone who inspires others to grow.

What Are the Key Leadership Skills Everyone Must Develop?

There are certainly different styles of leadership, but it's fair to say that great leaders generally share certain qualities. As I've said, effective leadership encompasses many of the skills we've already covered in this book—so here, we'll run through essential leadership skills that we haven't yet focused on. This isn't an exhaustive list—rather, it's the qualities that I'd prioritize if I were looking to bring out the best in people.

Those qualities are:

- Motivating others
- Recognizing and fostering potential
- Inspiring trust
- Taking on and giving up responsibility
- Strategic thinking and planning
- Setting goals and expectations for everyone
- Giving (and receiving) feedback
- Team building
- Positivity
- Authenticity

Let's explore each one in turn.

Motivating others

The most impactful thing any leader can do is motivate others to be the best they can be. It's essential if you want people to succeed and, in turn, realize the company's vision. That's the big-picture responsibility of any leader. But, let's be honest, there's a more mundane, everyday side to motivation, and that is simply getting people to do what you need them to do. This can be a big challenge for some leaders, especially if you aren't used to handing over responsibility for tasks. (I talk more about delegation later in the chapter.) The good news is that people generally want the same thing from work—they want clarity, they want to feel their work matters, and they want recognition for a job well done. If you can give them that, your life as a leader will be much, much easier.

Here's how to do it:

- Make sure people know how their role contributes to the company's vision, so they understand the value of their work.

- When it comes to specific tasks, be clear on what's needed, why it's needed, and when it's needed.

- Give people the autonomy to accomplish the task in their way. If you've ever been micromanaged, you'll know all too well the negative impact it has on motivation.

- Show your appreciation and celebrate success. Individual feedback, public praise, and team celebrations are great ways to motivate people.

Recognizing and fostering potential

Great leaders look for potential, not performance. They can spot potential a mile away, and give people the opportunities to achieve their potential. Of course, every team member is different, so there's no one template for potential. Rather, getting the best out of people means playing to their individual strengths. Here's what that may look like in practice:

- Don't fall into the trap of getting people to think and act like you. If you want to leverage people's individual strengths, you have to encourage them to think and act like *them*. Allowing people to be their authentic selves is a great way to unleash their potential.

- Help people develop their critical thinking and decision-making skills (see Chapters 5 and 6), for example, by encouraging them to think about the likely outcomes of certain decisions and the impact on various stakeholders.

- Encourage people to take risks, step outside their comfort zone, and test new ideas. (For this to really work, people need to know that it's okay to fail sometimes.)

- Build a high-performance team (more on building a team later in the chapter). Because, when you surround people with high performers, you're more likely to strengthen their potential.

- Don't let people grow complacent—constantly encourage them to develop their skills and think about the next stage of their career, whatever that may be.

Inspiring trust

Without trust, how can you inspire anyone to do more and become more? It takes a lot to gain people's trust—and keep it—but it's a vital part of leading others, especially in times of business disruption, change, and uncertainty.

So, what makes a leader trustworthy? The following behaviors are a good start:

- Being ethical (circle back to Chapter 14), which encompasses being honest and transparent, keeping promises, and generally making sure you don't say one thing and then do another

- Making your values clear and, of course, living those values

- Standing up for what you believe in, even when it feels like you're swimming against the tide

- Being an excellent listener, which shows you care (People who care can generally be trusted to do the right thing and act in the interests of others.)

- Facing difficult issues head on, rather than pretending everything is fine

Taking on and giving up responsibility

Read anything about leadership and you'll see a lot about delegation. But I like to think of responsibility in two halves—sure, there's delegating responsibility, but good leaders also *take on* responsibility. When it comes to the latter, leadership means:

- Knowing what to delegate and what *not* to delegate

- Being willing to take charge, in good times and the not-so-good times

- Being accountable and willing to take the blame when things go wrong

- Building a reputation as someone others can count on

In terms of delegating effectively:

- Remember, delegating gives others the chance to grow their skills. If you refuse to give others responsibility, you deny them learning opportunities.

- Play to the strengths of your team and allocate responsibility accordingly. Who has the specific skillset to achieve the result you want?

- Once again, clarify the desired end result and then give people the freedom and autonomy to decide how best to do the work. Be open to new ideas and new ways of doing things—your way isn't the only way.

- Ensure people have the knowledge, resources, and tools they need to succeed.

- Decide how you'll monitor progress without micromanaging. For example, you can agree how the person will report back to you, and how often—as well as the best way for them to raise any questions.

- Provide regular feedback, both so you can give praise and so you can take action when things aren't going according to plan. There's more on giving (and receiving) feedback later in the chapter.

- Revisit the pointers on motivating others from earlier in the chapter because these will also help you delegate more effectively.

Thinking strategically

In Chapter 5 I talked about the importance of critical thinking at all levels of a business, especially in this fast-paced world where we're constantly bombarded with information. Strategic thinking means applying these critical thinking skills in order to take a wider view, solve business problems, and make a long-term plan for the future.

This is what leaders do. They take a holistic view of things, rather than focusing in on the here-and-now or thinking only of their own job and responsibilities. Given the rapidly changing nature of work, I'd argue this is an important skill for everyone to cultivate, not just those at the top of the chain.

To boost your strategic skills, you can:

- Keep in mind the difference between *urgent* and *important*. Urgent fire-fighting tasks can suck up a lot of your time and energy, leaving very little bandwidth for those things that are important from a big-picture perspective, but not urgent. Combat this by constantly reminding yourself of your priorities, and managing your time accordingly (turn to Chapter 17 for more on time management).

- Free up time for strategic thinking by delegating effectively. It's hard to see the bigger picture when you're bogged down in the day-to-day minutia.

- Ask better questions—the kinds of strategic questions that help you spot new opportunities, work out how to respond to challenges, and think a few steps ahead. A good example is asking, "Where will our growth come from in the next few years?"

- Harness your critical thinking skills (Chapter 5) to gather data and find solutions to your strategic questions. Don't answer strategic questions based on assumptions or gut instincts.

- Look for connections and patterns that others might not see.

- Don't be afraid to take risks that may not ultimately pay off. To succeed, you have to be willing to fail. I get that this makes many people uncomfortable, but it's an important part of striving for bigger things. (Circle back to Google's "20-percent rule" from Chapter 8.)

- Include others in the strategic process. As we saw in Chapter 13, diversity of thought is a good thing.

Setting goals and expectations for everyone

Setting goals is a great way to drive performance. Traditionally, this is done in a top-down way, with leadership setting strategic and management goals, and managers setting goals for teams and individuals—a process that may only take place once a year. But I suggest you consider a much more dynamic approach known as objectives and key results (OKRs). This technique is used by Google (among others) as a goal-setting framework.

I urge you to read up on OKRs in more detail, but as a very brief summary:

- OKRs consist of short, inspirational objectives and, typically, two to five key results (measurable deliverables) for each objective. In other words, the objectives define where you want to go, and the key results measure your progress towards those goals.

- Individual and team OKRs don't cascade down from the top. Rather, leadership sets some strategic OKRs for the business, then each team and individual designs their own OKRs that contribute to achieving the company's strategic OKRs.

- What I love about OKRs is they encourage collaboration. OKRs make it easy to understand how everyone in the organization has a critical role to play in achieving the strategic OKRs. Everyone is moving towards a common objective.

- OKRs work best when they are simple and agile. Forget annual goal-setting; OKRs are typically set on a monthly or quarterly basis, meaning the business can stay nimble and respond to change.

- OKRs are not a tool for evaluating employees. If people are going to set ambitious OKRs (which is what you want, rather than setting goals that are too easy), they need to know they won't be negatively impacted if they don't achieve every single OKR. A good rule of thumb is that it's fine to achieve 60–70 percent of OKRs.

- OKRs are intended to be lightweight, so don't overburden the process with lots of meetings or documentation.

Giving (and receiving) feedback

One of the most important things a leader can do is support their team and help people perform better—which relies on the ability to give feedback, both positive and negative. As humans, we often have a tendency to look for the things that are not going well, so that we can correct them. But it's just as important to focus on what people are doing well, and celebrate that success regularly.

That said, giving someone positive feedback is fairly easy. It's generally the negative feedback that leaders struggle with. Here are a few tips for giving negative (or as I prefer to call it, *constructive*) feedback:

- Don't put it off. Otherwise, you may end up overwhelming the person with a very long list of their faults or—worse—blurting out negative comments in a moment of frustration or anger. Have a process in place for regular catch-ups, ideally weekly or at the very least monthly, where you can chat through progress, give feedback, answer questions, and (if you're using them) check in with OKRs.

- Do it in private. Public praise can be great for morale, but constructive feedback should always be given behind closed doors.

- Be specific, not emotional. Don't be awkward, apologetic, or aggressive. Just treat it as a straightforward conversation, using specific, concrete examples instead of opinions or emotions.

- Don't dilute constructive feedback with praise. If you sandwich negative comments with a positive comment either side, there's a risk the person may only hear the good stuff. So, while it's important to regularly give people praise, I wouldn't do it at the same time as constructive feedback.

- Ask questions like "What was your thought process for X?" or "What do you think could be done better next time?" This encourages self-awareness and critical thinking, and can help you identify any underlying issues, lack of understanding, or miscommunications that need to be rectified.

- Agree a way forward, based on the positive results and behaviors that you want to see in future (as opposed to you telling them exactly what to do). Check in against these goals in your regular catch-ups.

Of course, feedback is a two-way street. So as well as giving feedback, any leader must also invite and accept feedback. However, this may not always be positive. Here are some tips for coping with constructive feedback from others:

- Don't react straight away. Take time to reflect on what you've been told, giving yourself an honest (non-emotional) evaluation. Remember, constructive feedback is an opportunity to improve.

- Ask yourself, "How could I do this better in future?" Identify key learning points and actions that will help you improve.

- Check in on your own progress—through self-evaluation and asking others for feedback.

Building a strong team

I talked about teamwork and collaboration in Chapter 9, but leaders must also be able to build teams—much like a football manager picks strong players who perform different roles and then shapes those players into a cohesive team unit. In my experience, some leaders actively shy away from picking strong people who may outshine them, but this isn't a smart move for the long term. A strong team makes its leader look good. An underperforming team doesn't.

Assuming you're involved in the selection process (you may not be, for example, if you've "inherited" a team), you obviously want to recruit the right people. This means choosing people with the right skills, of course, but don't overlook the importance of diversity—diversity of skills, thought, background, age, gender, culture, and so on. (See the importance of diversity, Chapter 13.) Also look for people who are good team players, who enjoy working towards a common goal.

With the right people in place, it's up to you to mold them into a team:

- Foster that sense of team spirit by building connections within the team. People will take their cue from you, so model the behaviors you want to see: connecting as human beings, showing an interest, listening to each other, treating people with respect and dignity, and supporting one another.

- Remember, it's a team of individuals. Each person will bring their own unique skills and experiences, be motivated by different things, have different working styles, and so on. Embrace this, and try to see this diversity as the asset it is, rather than trying to force everyone to behave the same way.

- Set your expectations. We've talked about driving performance already, but you'll also want to set your expectations for how the team functions. For example, do you want to create the sort of team where everyone plays a role in the decision-making? (Ideally, yes.) Then make that clear from the outset.

- Give feedback and reward a job well done, if not through financial bonuses, then with gratitude. If appropriate, you can also hand over additional responsibilities, thereby showing trust in your team.

Positivity

The attitude you bring to work has a huge impact on the people around you. If you show up with a negative "this won't work, that thing sucks, why do we bother" kind of attitude, it'll spread throughout your team. The good news is, anyone can cultivate a positive attitude, even if you're a natural pessimist. I'm not talking about blind optimism here or pretending that everything is always fine. Rather, I'm talking about acknowledging the positives and not spiraling into negativity every time something goes wrong.

Here's how you can lead from a place of positivity:

- Think carefully about the language you use, verbally and in writing. Use words with positive connotations—turning a "problem" into an "opportunity" being a prime example.

- I've said it before but I'll say it again: celebrate successes, big and small. Highlighting the little wins on a regular basis can be just as impactful as sporadically celebrating the big wins.

- When things don't go according to plan, stay calm. Try to respond from a place of kindness, patience, and empathy.

- Resist the urge to complain to others on your team. There's a scene in *Saving Private Ryan* where Tom Hanks, who plays a captain, is asked by his team of soldiers why he never complains. His answer is worth keeping in mind: "I don't gripe to you. . . . Gripes go up, not down. Always up."

- Inject some fun into the working week. Whether it's through team lunches, away days, weekly bowling nights, casual Fridays, or occasional after-work drinks, fun activities boost morale and help the team mesh together.

Authenticity

In the previous chapter we saw that unethical behavior is often inauthentic behavior—where a brand says they behave one way, then acts another. But it's not just companies who need to practice authenticity—we need authentic leaders, too. And being an authentic leader means being able to connect with people on a human level. For me, it's a key part of building trust.

But what makes someone authentic? Some of the attributes frequently associated with authentic leaders are:

- Leading with empathy, or leading "from the heart"

- Being honest, open, and transparent

- Having a strong ethical and moral compass

- Being self-aware—meaning a good leader is aware of their weaknesses as well as their strengths, and open about those weaknesses

- Learning from mistakes

- Bringing your whole self to work, as opposed to having one persona for work and one outside of work

How You Can Become a Better Leader

I've covered practical tips for each of the skills mentioned in this chapter, but let me round off with some general learning points to help you apply these skills and strengthen your leadership muscles:

- Seek out learning opportunities. This could be signing up for a free online course, finding a mentor, reading books about leadership, or whatever. As a starting point, I highly recommend you read the Simon Sinek book I mentioned, *Leaders Eat Last*.

- Think about a leader you admire or have admired in the past—this can be someone you've worked for, or a leader in the public eye.

Note what exactly it is you admire about them and consider how you could build those qualities in yourself.

- Ask for more responsibility. Tell your manager that you're keen to grow your leadership skills and ask if you can take some of the load off their plate.

- Remember to think beyond your individual job and responsibilities. Good leaders take a holistic view, so use your initiative and think about the challenges and opportunities that your team and the wider business might face.

Key Takeaways

Here's a reminder of the key points on leadership:

- With the changing nature of work—distributed teams, increasing diversity, humans transitioning to more creative tasks, the gig economy, fluid organizational structures, and so on—leadership skills will be important not just for those in traditional leadership roles. Increasingly, more people throughout the organization will be required to lead, whether they're leading a project or an entire department.

- Most of the chapters in this book can be classed as a leadership skill, so as you circle back to chapters, think about how each skill could help you become a better leader.

- Other skills to develop include motivating others, recognizing and fostering potential, inspiring trust, taking on and giving up responsibility, strategic thinking and planning, setting goals and expectations for everyone, giving (and receiving) feedback, team building, positivity, and authenticity.

- Although leadership is a collection of different skills (and each leader will have their own unique leadership style) ultimately good leadership is about helping other people grow.

Building a strong reputation is part of being a respected, trusted leader. In the next chapter, we'll delve into the notion of reputation and personal brand—and how it can help you develop the career of your dreams.

Note

1. The determinants of leadership role occupancy: Genetic and personality factors; The Leadership Quarterly; https://www.sciencedirect.com/science/article/pii/S1048984305001232

CHAPTER 16
BRAND OF "YOU" AND NETWORKING

Back when I was starting my career at the University of Cambridge, if someone had told me that I'd end up with more than two million social media followers, and regularly doing live streams online, I'd never have believed them. I just wanted to build a good reputation in my field; little did I know that would grow into a "brand."

Today, building and maintaining my personal brand is an important part of my job. But it's becoming important in so many professions, way beyond the realms of influencers, entrepreneurs, and gurus. In this digital age, your reputation—everyone's reputation, really—exists in the online world as much as the offline world. You can take advantage of this to establish your expertise, network and make new connections, grow your career (or, if you're self-employed, generate new business), and, ultimately, build your own personal brand. Whether you're employed, self-employed, a leader in your industry, or someone who's just starting out in their career, being your own brand ambassador may just be the most important job you'll ever have.

What Is the Brand of You?

You're already familiar with the concept of a brand. In the traditional sense, a brand is the culmination of all those things that make a company stand out, such as its values, visuals, and tone of voice. Basically, anything that contributes to how people perceive a business, that's its brand. As Jeff Bezos neatly puts it, "Your brand is what people say about you when you're not in the room."

Increasingly, this notion of brand is being attached to individuals, giving rise to the term "personal brand." Just as with a company's brand, your personal brand is the culmination of all those things—skills, experience, personality—that make you *you*. It's your reputation. It's what others might say about you when you're not around.

Whether you like it or not, you already have a personal brand. Everyone does. If I were to Google your name, what impression would I get of you based on the first things to come up (typically your LinkedIn and social media profiles)? That's your brand. Personal branding simply means taking control of that online reputation and shaping it so people see you in the way that you want to be seen. Do this well, and you can establish yourself as an expert in your chosen career, among people who have never even met you, and attract exciting new opportunities your way, by making sure you're at the forefront of people's minds when they think of your job or industry.

In short, your personal brand can help you stand out from the crowd, whether you're an architect, entrepreneur, designer, blogger, lawyer, or whatever.

This doesn't mean you seek to become world-famous. Sure, there are household names who have turned themselves into wildly successful brands, such as Oprah Winfrey (personal brand: helping people live their best lives) or Richard Branson (personal brand: visionary entrepreneur, risk-taker, daredevil). But there are also people like me—not a household name, but very visible in my field—who have built highly successful brands in various niches.

So Google my name and the first things you'll see are my own website, then my latest tweets, then my LinkedIn profile, then my YouTube channel, and then my other social media profiles. Even just a quick glance at these results is enough to tell you I'm an expert in future technologies, digital transformation, and driving business performance. You'll see the same (professional) photos of me, and read the same voice (mine). All of that contributes to my brand. It's consistent. It tells a story about who I am and what I do.

Now, I spend a lot of time on my brand, so don't be alarmed thinking you need to go to all these lengths—you don't. As we'll see later in the chapter, you can cherry-pick your preferred methods, start small, and build from there. For example, you might start by polishing up your LinkedIn profile, adding testimonials from others to your profile, inviting new people to connect with you, and perhaps eventually publishing your own LinkedIn articles in your field of expertise. Then you might build on that by sharing more content on Instagram or Twitter.

The point is to take control of your reputation and use these tools to your advantage, whatever that means to you.

You'll notice that most of this chapter is devoted to personal brand, as opposed to traditional networking approaches like going to local networking meetings. That's not to say in-person networking is no longer important; it's just that, in this digital-first age—and particularly with global teams and more remote or hybrid working—the emphasis is definitely tipping towards building and maintaining connections through digital channels. Sharpening your online reputation will help you do this.

Why Does Personal Brand Matter?

Getting hired used to involve looking for vacancies, submitting an application or CV, going for an interview, and then hoping your current employer would give you a good reference. But that is changing. Your next job interview could be a video call. You may never meet your manager or project lead in person. You may be part of the burgeoning gig economy and regularly scouting for new projects (see Chapter 11). And, importantly, as the talent pool is now global, you could be up against people from all over the world. There's also rising interest in recruiting "passive" candidates, where employers reach out directly to qualified people who aren't currently seeking a job. All of this means you need to stand out online.

In this way, building your personal brand can help you secure your next exciting job, or bring new clients your way if you're self-employed. But what if you're happily employed and not looking to jump ship anytime soon? Reputation still matters. Reputation is what helps you define your skills and knowledge, inspire trust, enhance your job security, expand your network, and progress your career. Essentially, having a strong personal brand is about *bringing opportunities your way*, whether it's getting a promotion, winning new clients, pulling in job offers, writing a book, or being invited to speak at next year's convention. It really does pay off. My very first book came about because a publisher saw I was writing a lot of content and sharing it online.

If all that hasn't inspired you to work on your personal brand, perhaps these sobering stats will: 70 percent of employers research candidates on social media during the hiring process, and 57 percent have chosen to *not* hire someone because of what they found online.[1] Your online reputation is therefore incredibly important. Interestingly, the same survey also asked employers what kinds of positive content led them to hire someone. The top results were:

- The candidate's profiles and information supported their qualifications for the job.

- The candidate showed creativity.

- The candidate's content conveyed a professional image.

For me, this perfectly illustrates the importance of curating your online reputation, so that it shows you as the knowledgeable, creative, thoughtful, authentic person you are.

The benefits of personal branding extend to employers, too. Having employees with strong personal brands help to give the business more exposure, improve your employer brand, and attract more talent your way. If you're an employer, it's well worth encouraging your people to invest time and energy in their personal brand.

How to Boost Your Personal Brand and Expand Your Online Network

Building an effective personal brand takes a bit of work, but anyone can do it. And the good news is, you can tailor your approach to suit you. In other words, please don't view the following tips as a cookie-cutter approach that you have to follow step by step. At the end of the day, personal branding is just about crafting your digital reputation—and the best way to do that will depend on your industry, your strengths, and your audience.

Let's begin with some general tips:

- Find your niche (then widen around the edges). For most people, establishing yourself in a niche topic is the best way to begin building your brand. I started by focusing my brand on strategic performance management, for example, then expanded into artificial intelligence and wider future trends. If you're an accountant, you might start by focusing your brand on business finance, then expand into business strategy and leadership. A photographer might establish themselves as expert in food photography, then expand into entrepreneurship and helping others build a successful photography business. You get the idea: your brand can evolve over time.

- Think strategically. I've said it a few times in this book, but look beyond your own current job responsibilities. Try to take a longer-term view of your job or industry and identify the areas that will grow in importance over the coming years. One of those could be your niche, or an area for you to expand into in future.

- Write a personal brand statement. Early on, it's a good idea to identify exactly what you want your brand to be (keeping in mind that it may evolve over time). Distill this down into a short, snappy description of one or two sentences—your own personal tagline, if you like. For example, I'd describe myself as a world-renowned futurist, influencer, and thought leader in the fields of business and technology, with a passion for using technology for the good of humanity. How would you describe what you do and for whom? You don't have to share this statement publicly, although it might inform your online bios.

- Consider adopting a brand name. Some people take personal branding a step further by giving their brand a name, like the Data Guy (tech and data expert) or Money-Saving Expert (personal finance tips). This may not be right for you—personally, I prefer to use my own name—but it may be worth considering.

- Start with small steps. Early on, everyone was telling me I should be doing YouTube videos, that YouTube was the place to be. But I just wasn't ready to take that step. I much preferred to focus on written content for my website, Twitter, and LinkedIn. Nowadays, I'm very comfortable doing livestreams and YouTube videos every week, but it took me years to get to that point. So don't put pressure on yourself to be on all the platforms at once or to adopt multiple content mediums. You can start small just by doing a bit more on your preferred social media platforms.

When it comes to using social media effectively:

- Have a professional up-to-date photo in your social media profiles, and use the same photo across different platforms to ensure consistency.

- Clean up your profiles by deleting any content that you wouldn't want potential employers to see.

- If you want to keep one of your social media profiles strictly for personal content only, make sure you set your privacy controls to the max so that your profile can't be found in searches. For example, you might keep your Facebook or TikTok profile private if you want a space to share more personal content (although, as we saw in Chapter 4, always think carefully about sharing personal details on social media). Then you can focus your personal brand on other platforms like LinkedIn and Instagram, keeping those profiles public so anyone can see them.

- Be yourself. While you want to cultivate a professional brand, it's important to let your personality shine through in your social media posts. Write in the way you'd normally speak. Be authentic. Be honest. Talk about things that really matter to you (rather than trying to hop on the latest trends). And don't pretend to be someone you're not. This is all part of ensuring your brand stays *consistent.*

- Share what you're learning. Something that I've found impactful—and easy!—is sharing interesting and relevant news stories from my industry on social media. This really helped me build my profile, and stay knowledgeable on what's happening in my field. To keep up to date with interesting and relevant news stories, you can subscribe to industry newsletters or, even easier, set up Google alerts for certain keyword topics. Do be sure to add your own message when you share something on social media—even if it's just "I came across this today and thought I'd share it. What do you guys think?" And do apply critical thinking (Chapter 5) to ensure you're sharing content from reputable sources.

- Join industry groups on social media platforms. Then make yourself known by engaging with posts, answering questions, and liking, commenting, and sharing other people's content in the group.

- Be generous with your time and knowledge. Basically, be helpful to others online by responding to questions and comments, and generally engaging with them. And do take the time to like or amplify other content that you found engaging, inspiring, or useful. It's all about being reciprocal.

- Make new contacts as often as you can, especially on LinkedIn. You can do this by identifying people you want to connect with in your field and sending a certain number of invites each week, with a short personal message. Make a habit of this and your network will soon grow.

- Create polls on LinkedIn, Twitter, and Instagram. This is a great way to pose interesting questions and boost engagement. You can always mix it up by posting a mixture of professional and more general questions.

- Post quality photos and videos from your work life. People love visual content, so if you're at a work conference, attending an industry event, on the way to visit a client, or whatever, share it. You can mix it up with occasional "everyday" photos and videos while still keeping it fairly professional (think your morning cup of coffee when you're working from home, the view on your morning run, that sort of thing).

- Really, you can post any sort of content that will help to cement your reputation—it could be advice, thought-provoking questions, excerpts from presentations you've given, pro tips, how-to content, or whatever.

- If you really want to establish your expertise, consider writing longer-form articles and sharing them on LinkedIn.

- If you're ever stuck for something to share, inspirational quotes always go down well.

- Use cross-platform tools to make your life easier. For example, you can use a tool like Hootsuite to schedule your posts in advance and share posts across multiple platforms, such as Instagram and Twitter, all from one place. This means you can get maximum value from each piece of content, without having to physically post it in multiple places. The tool takes care of it for you.

- Try penciling in a specific time each day or week for social media. Building your brand doesn't have to be a full-time job, and you may actively want to limit the amount of time you spend on social media (it can be a huge time suck). So I find it helps to schedule posts in advance (see above), and block out specific times to check in with social media, reply to comments, and see other people's posts.

Looking beyond social media:

- Build a reputation within your *industry*, not just your current employer. For example, you might volunteer for industry steering groups, join cross-company committees and projects, attend conferences, and so on.

- Consider investing in your own website domain (either in your personal name or brand name). Only you can decide whether this is worth it, but I've certainly found it valuable, especially as someone who's in the gig economy and as someone who creates a lot of content.

- Explore other opportunities to share your knowledge, such as speaking engagements, writing a newsletter, or even writing a book—anything that helps establish your reputation as an expert.

Finally, while most of this chapter is aimed at individuals, it's really important that organizations also consider this new world of personal branding and what it may mean for their employees. A good starting point is to have clear guidelines on what is and isn't acceptable on social media. Beyond that, I recommend encouraging your people to share professional content online, talk about their work life and share positive work stories, and generally help build your brand online.

Key Takeaways

In this chapter we've learned:

- Your personal brand is the culmination of all those things—skills, experience, personality—that make you *you*. It's your reputation, essentially.

- Whatever your profession, having a strong personal brand can help you establish yourself as an expert, stand out from the crowd, and bring new opportunities your way—opportunities like getting a promotion, pulling in job offers, or winning new clients.

- The overwhelming majority of employers research potential candidates on social media. Yet another reason to carefully hone your online reputation.

- Anyone can strengthen their personal brand. It's just about crafting your digital reputation so others see you how you want to be seen. You can do this in very simple terms just by being a little more strategic with your social media posts, or you can go as far as publishing content and videos and having your own website.

- Good first steps are to find your niche, write your own tagline that defines your personal brand, then begin to improve your social media game.

Building your personal brand can seem like yet another thing on your to-do list, so it's really important you have strategies in place to manage your time and be productive, while maintaining a healthy work–life balance. Let's delve into time management in more detail and see what this looks like in our increasingly fast-paced world.

Note

1. More Than Half of Employers Have Found Content on Social Media That Caused Them NOT to Hire a Candidate, According to Recent Career-Builder Survey; PR Newswire; https://www.prnewswire.com/news-releases/more-than-half-of-employers-have-found-content-on-social-media-that-caused-them-not-to-hire-a-candidate-according-to-recent-careerbuilder-survey-300694437.html

CHAPTER 17
TIME MANAGEMENT

It's hard to know how the pandemic and the associated shift to more remote and hybrid working will affect work–life balance and stress levels in the long term. I know people who find working from home to be stressful and less productive, with the distractions of family life (not to mention that feeling that they can never really "switch off" from work at the end of the day). For others, it's a big improvement on the demands of traveling to and from the office every day, allowing them to focus more on getting the actual work done. I'll be interested to see future studies on the link between remote working, productivity, stress, and time management.

What's clear now is that time management is just as important as ever. Whether you work from home, work full-time in an office, run your own business, or work for an organization, the ability to manage your time effectively is essential for your workplace performance and, frankly, your mental health (see Chapter 20 for more on looking after yourself). It's especially important in this age of fast-paced work, information overload, and constant distractions from email and app notifications.

What Is Time Management?

Time management is the ability to use your time efficiently and productively, especially in a work context. Achieving this in practice often requires planning your time and thinking strategically about how best to spend your time—but more on that coming up later in the chapter.

The productivity myth

As we saw in Chapter 11, the traditional nine-to-five, five-days-a-week working pattern doesn't exactly reflect our capacity for productivity, because the average person is productive for less than three hours a day. In other words, that colleague who stays later than everyone else isn't necessarily getting more work done (probably the opposite is true).

Recognizing this, time management is about working *smarter* rather than working harder or longer. Someone who is great at managing their time will know when they're at their most productive and use that time wisely, reserving the less-productive hours for other tasks (or nonwork passions). In this way, time management is all about creating a better work–life balance. After all, your time is a precious—and finite—resource.

Some companies have really taken this to heart and are starting to introduce four-day workweeks. Unilever is one such company. In New Zealand, Unilever is trialing a radical approach where employees work four days a week but get paid for five days. The approach is based around the 100:80:100 logic, in which people keep 100 percent of their salary, work 80 percent of the time, and still deliver 100 percent of their output. If the trial is successful, Unilever says it will extend the initiative to other offices around the world.[1]

Even entire countries are transitioning to a four-day workweek as standard. Iceland trialed this approach between 2015 and 2019 and dubbed the trial an "overwhelming success." Today, 86 percent of Iceland's workforce is already working fewer hours (without taking a pay cut), or will be entitled to do so.[2] Best of all, productivity levels at the companies involved in the trial either stayed the same or improved—showing that productivity isn't about how much time you spend working, but how you spend that time.

Overcoming procrastination

There's often a conflict at the center of time management—and that conflict is all about self-control and motivation. Think of it as having an angel on one shoulder and a devil on the other. The angel motivates you to get a certain task done, while the devil is constantly coming up with reasons to put it off. Managing your time effectively is a question of which one is most persuasive—the angel or the devil. And this may be influenced by various internal and external factors.

Of course, we all know deep down that time management is about having the self-control to get things done and not to procrastinate. But the reality isn't that simple, since most of us are still guilty of procrastinating at some point or other. Why is procrastination so tempting? Why do we listen to that voice telling us that it's okay to put something off, even when we know it doesn't serve us?

Research shows that the various reasons why we procrastinate fall into two camps: *demotivating factors* and *hindering factors*.[3] Examples of demotivating factors might be fear of failure, anxiety, perfectionism, or simply not wanting to do the task because it's unpleasant. And hindering factors might include things like being exhausted, having goals that are too vague, or when the reward for the task is too far in the future. These are hindering factors because they literally hinder our motivation. And when our motivation is weakened, it's more likely to be outweighed or overpowered by any demotivating factors. As a result, the balance tips in favor of procrastination.

This is a simplistic summary—the psychological mechanisms behind motivation and the reasons for procrastination are obviously more complex and varied than this. But it goes to show that time management is often a case of making sure your motivation—which may or may not be hindered by external factors—outweighs any demotivating factors. The good news is there are plenty of practical strategies to keep your motivation up, avoid procrastination, and make the most of your time. More on this coming up later in the chapter.

Why Time Management Matters Now More Than Ever

I don't think I need to work hard to sell you on the benefits of time management. You're probably well aware of how good it feels when you manage your time efficiently, and how stressful it can feel when you don't.

Perhaps the best way to illustrate the importance of time management is to talk about what happens when we get it wrong. Some of the results of poor time management include:

- Procrastination (there's that word again)

- Inefficiency, lower productivity, and/or poor-quality work (often because we end up rushing in the end)

- More stress (especially when you feel you're not in control of your time)

- Poor work–life balance (because work tasks, and stress, eat into precious nonwork time)

- Missed deadlines (because we've simply run out of time or underestimated the time needed for a task)

- Negative impact on your professional reputation (see personal brand, Chapter 16)

On the flip side, if you can manage your time well, you can expect the opposite effects: eliminating (or reducing) the tendency to procrastinate, being more efficient and productive with your time, feeling in control and lowering your stress levels, having more time for nonwork passions, nailing your deadlines, and strengthening your reputation as a person that can be trusted to get the work done on time and to a high standard. Remember those lucky Icelanders who had transitioned to working fewer hours, for the same salary, while still getting the same (or more) work done? Workers involved in the trials reported feeling less stressed and at lower risk of burnout, and that their health and work–life balance was better. And they had more time to indulge in family life and hobbies.

Bottom line, time management helps you work smarter, so that you get the best out of your working life and, you know, *life* life.

It's no wonder that good time management is one of those soft skills that's perennially included on lists of most important or most desirable skills. But that's not the reason I've included it in this book. I've included it because it feels as though the nature of modern work and life is almost setting us up to fail from a time management perspective. We have apps that constantly ping with notifications, actively trying to draw us in so we spend more and more time on them (see digital addiction, Chapter 4). We have the pressures of juggling work and family life—which, especially for many women, can mean a huge amount of behind-the-scenes emotional labor to keep family life ticking along smoothly. We have emails flying into the inbox long into the evening, leaving many people feeling like work is constantly expanding into nonwork time. Life just feels faster and more demanding.

All this means we're at risk of those demotivating and hindering factors I mentioned earlier outweighing our motivation to get things done. Which is why we all need practical strategies to manage our time efficiently.

How to Boost Your Time Management Skills

Let's explore how individuals and organizations can enhance their time management skills and work smarter:

For individuals

- Do the important jobs first. People often like to get the most unpleasant task ticked off the list first, just to get it done. (As the Mark Twain saying goes, "If it's your job to eat a frog, it's best to do it first thing in the morning. And if it's your job to eat two frogs, it's best to eat the biggest one first.") Others like to get quick and easy tasks done first, just to feel like they're achieving stuff. But it's far better to prioritize in order of *importance*, regardless of whether or not it's hard. This brings me to the next point.

- Ruthlessly prioritize your time. I like to use the ABC method to plan my day and prioritize tasks in order of importance. An "A" task is my most important, must-do item for the day (or, if there's more than one A task, I label them A1, A2, and so on). "B" tasks are secondary tasks that are less important than A tasks—you never move onto a B task while there are still A tasks on the list. And C tasks are those that are nice to get done, but it's not a big deal if they don't happen that day. I start every morning with this method (or you could do it at the end of each day, ready for the next day).

- Set a time limit for each task. Once I've made my to-do list for the day, I set time limits for each task on the list. This ensures I don't let tasks expand to fill more time than they really need, and it keeps my day manageable because I know what I can realistically achieve.

- Build in buffers between tasks. When planning your time, don't absolutely pack your schedule. You never know when something may take a bit longer than you think, or when something more important may crop up. For example, if you know a task should probably take one hour, block out an hour and 15 minutes, thereby leaving you a bit of wriggle room in your day.

- Schedule breaks. In Chapter 11 we saw that productive people take regular short breaks, and you should do the same. You may find it helpful to block out time for breaks in your daily schedule, or set an hourly reminder to take a five-minute breather.

- Frame tasks and goals within a wider context. One thing that I find motivational is to focus on how specific tasks or goals will help me fulfill a longer-term vision. Yes, we humans like short-term results and instant gratification, but it also helps to see how tasks feed into the bigger picture. Ask yourself questions like, "How will this help my overall career?" "How will this help me achieve my long-term goal of X, Y, or Z?" Or "How will this help my organization achieve its vision?"

- Keep it bite-size. If a task feels overwhelmingly large, break it down into manageable, actionable pieces. Then prioritize those chunks in order of importance as per the ABC method.

- Find your productive hours. Productive people don't fill every hour of their day—they know when they work best and they make sure they get the important stuff done during those hours. For you, this may be in the morning, it may be late at night when the kids are in bed, or somewhere in between. The key thing is to know when you work best, so you can block that time out for the most important tasks, and avoid filling it up with meetings or less important jobs, which are better suited to other times of the day.

- Don't multitask. Multitasking is the enemy of productivity because you can end up not doing anything properly. Remember, time management isn't about being super-busy or working extra hard; it's about being smart with your time and effort. That means giving one task at a time your full attention, and finishing that before moving on to the next item.

- Eliminate distractions. I love working from home, but I recognize that some people find it distracting. It certainly helps to turn off notifications on your phone, turn on your phone's "do not disturb mode" when you need to, and set boundaries for anyone sharing your space. (For example, by saying, "For the next hour, I really need to get my head down and concentrate," or "When my office door is closed, it means do not disturb.") The same tips also apply when you're in an office environment.

- Learn to say no. Saying no is an art form, and if you can master it, you'll feel much more in control of your time. Very often, it's not even a case of saying no but setting expectations for when you *can* do something—for example, by saying, "I can't do this until next week," or "The next free time I have available for that is Friday afternoon." If you do need to say no, explain your current workload and be firm and polite, without being overly apologetic.

- Delegate nonessential tasks where possible. If something doesn't have to be done by you, consider delegating (or outsourcing) it. Circle back to Chapter 15 for more on leadership and delegation.

- Weigh up the consequences of doing something versus not doing it. When all else fails, ask yourself, "What will happen if I don't get this done?" If the answer is "Er, not much," then it's probably not that important. But if you know there might be serious consequences if you put it off, that might give you the extra motivational nudge you need.

- Reward yourself for getting jobs done, perhaps with a short walk, a fancy coffee, or a few minutes on TikTok—whatever floats your boat.

- Understand that sometimes—just sometimes—procrastination can be a good thing. The urge to procrastinate might be telling you something (for example, that a task isn't that important to you, or that you're tired and need a break). So circle back to the demotivating and hindering factors from earlier in the chapter and try to identify *why* you feel the need to put something off. And sometimes, the mind just needs a bit of time to wander, imagine, and be creative (see Chapter 8)—and that's also a good thing.

- Finally, if you find yourself not wanting to do many of the tasks associated with your job, then maybe it's time to switch jobs! Seriously, ask yourself whether it's really the right job for you, because it's not "normal" to dislike your job or feel constantly demotivated.

For organizations

I highly recommend business leaders invest in time management training for their teams to instill good practice across the organization. But it's also important to recognize that productivity (and productive hours) will look very different from person to person, meaning you should give people the freedom and flexibility to work however and whenever they work best. You may even consider going as far as implementing a four-day workweek (while still paying people for five days), which has been shown to deliver benefits for work–life balance, productivity, and stress.

For managers, I recommend asking people within the team when they're most productive and focused, so that you can leave people alone during those times and, where possible, schedule things like catch-ups and team meetings for less focused hours.

Key Takeaways

To recap the key points on time management:

- Time management is the ability to use your time efficiently and productively, especially in a work context.

- The average person is productive for less than three hours a day. Therefore, good time management isn't about packing your schedule or working longer and harder than anyone else—quite the opposite. It's about working smarter so you have less stress and more time for nonwork passions.

- When it comes to improving your time management, some of the best things you can do are plan your day carefully, set time limits for tasks (with buffers), and ensure you relentlessly prioritize the most important tasks first. It's also important to take regular breaks, avoid multitasking, and recognize the underlying factors behind any urge to procrastinate.

While every chapter in this book is a vital future skill, we're about to move onto a subject that I'm especially passionate about: curiosity and continual learning.

Notes

1. Future workplace; Unilever; https://www.unilever.com/planet-and-society/future-of-work/future-workplace/

2. Four-day week 'an overwhelming success' in Iceland; BBC News; https://www.bbc.com/news/business-57724779

3. Why people procrastinate; Solving Procrastination; https://solvingprocrastination.com/why-people-procrastinate/

CHAPTER 18
CURIOSITY AND CONTINUAL LEARNING

If I were to pick just one key takeaway from this book—the one skill that I think *everyone* must cultivate—it would be curiosity and continual learning. Whatever your age, whatever your industry, if you can spark your curiosity (and, crucially, keep that spark alive), you'll be giving yourself the best chance of a successful, fulfilling life—workwise and otherwise. Being curious is what introduces us to new people, new information, and new experiences. It keeps life interesting and stops us getting stuck in a rut. As such, it's key to maintaining an active, healthy mind, which, as someone who's embracing middle age, I consider a definite priority!

And in a work context, curiosity and continual learning is fundamental to being able (and willing) to embrace change. It ensures your skills stay sharp, that you can keep up with the major transformations taking place in the fourth industrial revolution, and that you stay relevant.

Luckily, it's easier than ever to access learning materials. From audiobooks and podcasts to online courses, learning can be done on your terms. You just need to cultivate that desire to learn, that all-important curiosity.

What Do We Mean by Curiosity and Continual Learning?

Curiosity is the desire to learn and understand new things—whether it's understanding how something works, learning a new hobby, trying new foods, visiting new places, or whatever. This desire to learn fuels a journey of continual learning (also called lifelong learning), which is the ongoing, self-motivated pursuit of knowledge.

We're born into this world with an in-built curiosity. To babies and toddlers, everything is new and fascinating. And as anyone who has small (verbal) children knows, "why" is one of their favorite words. That's curiosity in action. Why is that thing like that? Why are you doing that? Why do I have to . . . ? Why, why, why?

At some point, most of us lose this constant desire to question everything. It starts at school, hampered by traditional education systems that generally value correct answers over questions. (This is backed up by research that shows the average 6- to 18-year-old student asks just one question per hour-long class per month—a stark contrast to preschool children's rate of questioning.[1])

This is a shame, because curiosity is a key ingredient in the learning process. Learning is just easier when you *want* to learn. The good news is, even if you feel you've lost that childlike curiosity—and you're not alone—there are many practical ways to reignite your curiosity and maintain a curious mind (more on that later in the chapter).

The habits of curious folk

Like young children, curious people ask lots of questions, and they're not shy about it. In fact, to a curious person, there's no such thing as a stupid question. They're not afraid of being seen as dumb, or saying "I don't know." Nor are they afraid of being wrong; they'd rather learn something new and interesting than be right all the time. They're generally good listeners too, and have the ability to listen without forming assumptions or hasty judgments.

Curious people have active rather than passive minds—they seek out new information and experiences, rather than accepting the world as it's presented to them. This means they read a lot, often on a broad range of topics, although they may also dive very deeply into topics that really excite them. As such, curious people are rarely bored. How can they be when there's always something new to learn?

Curious people are explorers, essentially—explorers of information, places, people, challenges, possibilities, and anything else that might expand their mind. As Albert Einstein once said, "I have no special talent, I am only passionately curious."

This is a fairly broad description of curious people, but I want to delve into two specific elements that I believe are essential for curiosity and continual learning. They are humility and a growth mindset. Let's start with humility.

Humility

Being humble—basically, to be free from pride and arrogance—is central to curiosity, because it tells us that we don't know everything there is to know. It tells us that we can stand to learn more, do more, and become more. Humility is often confused with a lack of confidence or self-belief, when in fact the opposite is true. Humble people recognize their strengths as well as their weaknesses, but they don't seek to hide their weaknesses, which are, after all, just opportunities to grow. This inner confidence is why a humble person has no fear of looking stupid or asking "stupid" questions—it's all part of growing.

Humility is an especially important quality—albeit not an immediately obvious one –for leaders and managers to cultivate (see Chapter 15, "Leadership Skills"). In his book *Good to Great*, Jim Collins identifies two traits that are common among CEOs of organizations that transitioned from average market performance to superior performance. Personal humility was one of those traits. (The other was indomitable will.) Think about it for a second; humble people are more likely to listen to what others have to say, to embrace feedback, and to work well with others—because humble people aren't under the illusion that they're the smartest person in the room.

We'll talk more about practical steps to ignite curiosity later in the chapter, but adopting a humble mindset is certainly an important first step. If you believe you have much to learn from others, you're more likely to feel that desire to learn.

Growth mindset

Psychologist Carol Dweck coined the phrase "growth mindset" in her groundbreaking book *Mindset: The New Psychology of Success*. If you haven't yet read this book, I highly recommend it as part of your continual learning journey. In *Mindset*, Dweck argues that success doesn't come from intelligence, talent, or education; it comes from having the right mindset—specifically, a *growth mindset*. This is backed up by her many years of research showing that the attitudes of students—in particular, their attitude to failures and setbacks—had a significant impact on their achievement.

Someone with a growth mindset believes they have the ability to grow, improve, and learn. They see obstacles, failures, and challenges as a chance to grow. And, importantly, they believe that, while everyone has inherent qualities and traits, success comes from constant personal development and continual learning. This is in contrast to someone with a *fixed mindset*, who believes they're limited by fixed, inherent traits and abilities that can't be changed or improved. Setbacks and failures, however small, can be devastating to someone with a fixed mindset, regardless of how talented or intelligent they are, because they believe that if they fail at something, it's because they lack the natural talent to succeed.

In the fixed mindset, you've either got it or you ain't. In the growth mindset, even the most basic abilities can be developed with hard work.

Of course, most people don't fall hard and fast into either a growth or fixed mindset. For most of us, we sit somewhere on the spectrum between the two, perhaps sometimes leaning more one way than the other.

Think about times when you've exhibited a fixed mindset. For example, maybe you've caught yourself saying something like, "Ugh, I can't do math." Adopting a growth mindset, you'd instead say, "I can't do math *yet*. But I can learn," because almost anything can be learned and improved with practice. I've noticed there's a lot of emphasis on this mindset—this notion that you don't know X, Y, or Z *yet*—in UK primary schools, which is interesting to see. As adults we can certainly all benefit from acknowledging our weaknesses or gaps in knowledge, without being limited by them—because we always have the ability to improve.

If humility will help you stay curious, having a growth mindset will help you commit to a journey of continual, lifelong learning. I'll talk about how to cultivate this mindset later in the chapter.

The darker side of curiosity

Before we move on, let me quickly acknowledge that curiosity doesn't always manifest as a positive desire to accumulate intellectual knowledge or to experience interesting new things. Someone who watches reality TV (yes, we've all done it) might be doing so because they're curious about the lives of the people on screen. Every tabloid story about a celebrity's love life is playing on readers' curiosity. It's easy to see how curiosity can tip over into gossip or nosiness (which is, after all, a desire to learn—it's just misdirected towards things that are unimportant or personal). So when I talk about asking lots of questions as a key trait of curious people, I'm not talking about prying.

Why Do Curiosity and Continual Learning Matter?

As Sir Ken Robinson, a key figure in education, put it, "Curiosity is the engine of achievement." In other words, it's a natural driving force that propels us forwards, on to success and personal fulfillment. For this reason alone, I believe we all have an obligation to stay curious. Where would we end up without that natural driving force? Stuck in a rut, probably. Bored, no doubt. Overtaken by changes in our job and industry, for sure.

If that isn't enough to convince you, here are some other reasons to prioritize curiosity and continual learning:

- It enables you to keep an active mind, which is really important for mental health and cognitive strength. I firmly believe curiosity keeps you mentally young.

- It widens the mind and opens you up to new topics, alternative opinions, and different cultures. (See Chapter 5, "Critical Thinking," and Chapter 13, "Cultural Intelligence and Diversity Consciousness.")

- It can be a springboard to doing your job better and more creatively (see Chapter 8, "Creativity"). In fact, curiosity is central to innovation. Most of the breakthroughs that have advanced humanity stem from someone's curiosity.

- It enables you to learn from mistakes. When curious people fail, they question *why* they failed, so they can do it better next time.

- It helps you build and maintain better interpersonal relationships (see Chapter 10, "Interpersonal Communication"), because curious people are interested, ask questions, and listen actively to what others have to say.

- It makes life more exciting! A curious life is very unlikely to be a boring life, because there's always something new to learn, explore, or improve.

All things considered, curiosity is important for success and satisfaction, not just in any job but in life in general. For me, it's the key to personal growth.

And for organizations, fostering curiosity is all part of driving business performance—in fact, curiosity is much more important to business performance than originally thought.[2] This is probably because curiosity helps people within the organization adapt to change, make better decisions, and come up with more inventive solutions.

How to Boost Your Curiosity and Embrace Lifelong Learning

Now it's time to explore practical ways for individuals and organizations to cultivate curiosity and lifelong learning.

For individuals

Let's start with humility, because that's one of the driving factors of curiosity:

- Be honest with yourself. Humility starts with truth, so think honestly about your weaknesses as well as your strengths. Be willing to admit your mistakes and take accountability, without relying on excuses. These mistakes or weaknesses show where you have room to grow.

- Practice acceptance. While you want to be honest about your shortcomings, don't be hard on yourself. Try to look at yourself without judgment and negativity—to accept yourself as you are, as the first step in learning to be or do better in future. Mindfulness techniques can give you some really useful acceptance skills.

- Practice active listening. Active listening enhances so many of the skills in this book and humility is no different. Listen intently to what others are saying and invite feedback at every opportunity. Importantly, listen with an open mind and leave your assumptions or preconceptions at the door.

- Ask for help. Recognizing when we need help—and being able to ask for that help—is an important part of humility. Likewise, don't be afraid of saying "I don't know." You don't have to be the smartest person in the room.

- Get comfortable with anxiety and uncertainty. Admitting that you need help, have failed at something, or don't know how to do something can be uncomfortable, as can any sort of change. Try to "sit with" these feelings when they crop up, rather than rushing to solve or unpack them. Again, mindfulness can be great for this, because it teaches you to acknowledge and accept thoughts without judgment.

Cultivating a growth mindset is key to continual learning:

- If you haven't already, do read Carol Dweck's book *Mindset*. In this chapter, I've only scratched the surface of her fascinating research.

- Think about where you currently sit on the growth mindset versus fixed mindset spectrum.

- Try to see challenges and failures as opportunities for self-development. It might help to think about a past challenge you've encountered, and how it ultimately led you to become stronger or better at something. After all, every athlete who's ever won a gold medal has no doubt had their fair share of injuries, losses, and setbacks along the way.

- Reward yourself for hard work. The growth mindset prioritizes effort and hard work over natural talent, so when you've worked hard at something—even if it hasn't been a total success—give yourself a mental pat on the back or a physical reward.

- Remember the power of *yet*, as in "I don't know how to do this *yet*." Notice how you talk about your own skills, or the talents of others, and reframe your language accordingly. For example, instead of saying, "She's so good at that," you could say, "She must have worked really hard to develop that skill."

- Be realistic. Learning any new skill takes hard work and patience. And that's okay. Embracing the journey is all part of the growth mindset.

- Practice humility (see above). Having a growth mindset means being willing to mess up, admit what we don't know (yet!), and expose our flaws as well as our strengths—all part of being humble.

And here are some general tips for developing your curiosity and continual learning:

- Make time for learning. I know life is busy, but do make time to read, watch, and listen to new things. Whenever you're learning a new skill, carve out time to practice it regularly.

- Do something different, today, tomorrow, the day after that. Just try something, anything, new. You could, for example, walk into a bookshop and pick up the first book that catches your eye. Or throw a dart at a map on the wall and cook a meal from that country's cuisine. Or listen to an entirely new genre of music. Or walk down a street you've never walked before, noticing the different buildings and gardens as you go.

- Keep an ideas journal. I've mentioned this before, but I find it really helpful to jot down ideas, quotes, and other interesting snippets in a journal. You could also use this to jot down questions that occur to you (so that you can research them later), or ideas for new activities to try, or even learning goals.

- Never say never (and never say boring). When you say you'll never want to do something, or declare something you've never tried is "boring," you're instantly closing yourself off to new possibilities. Curious people are open to all sorts of experiences.

- Make full use of the ever-expanding array of learning resources on offer. There are free online courses, live streams on Instagram and YouTube, podcasts, books, audiobooks (great for making the most of long journeys), and more.

- Ask open-ended questions. Try to emulate children's tendency to question everything, especially with those Who, What, Where, When, and Why questions. Pair this with deep listening.

- Learn not to take things at face value. (Revisit the tips for critical thinking from Chapter 5 for more on this.)

- Set yourself learning goals—learning a certain song on the piano by the end of the month, completing five hours of language learning each week, or gaining a project management qualification by the end of the year. Use the SMART goal-setting technique, and make sure your goals are Specific, Measurable, Achievable, Relevant (to your career or life goals), and Time-Specific.

- Don't direct curiosity towards gossip or unimportant details. It's a waste of your precious time and energy.

- When you see others succeed, first celebrate their success. But then engage your curious mind and ask them what it took to become so successful or good at what they do. What challenges did they overcome? What steps did they take on their journey?

I'd also like to say a few words about fostering creativity in children, because it's a topic that's close to my heart. If you have children (or work with children), do encourage them to ask questions, try new things, read books beyond what they're told to read at school, take up new hobbies, and so on. Praise their hard work rather than inherent abilities. (For example, say something like "You did so well on your spelling test. You must have worked really hard to learn those words," as opposed to "You did so well on your spelling test. You're so clever.") Also, remember the power of *yet* in conversations with your children; when they say they don't know or can't do something, add the word "yet" on the end. Keep doing it until they roll their eyes, then do it some more. And of course, do the same when you talk about your own skills and knowledge.

For organizations

It's really important that companies and leaders encourage curiosity and continual learning in their teams. A great way to do this is to celebrate learning and effort instead of just celebrating outcomes or output (for example, by linking performance evaluations to learning goals). As part of this, you can let employees broaden their interests and define their own learning goals. You want to encourage self-motivated learning, but the organization will no doubt need to invest in more formal learning and upskilling resources as well.

You can also hire for curiosity, by making curiosity one of the attributes you assess candidates for.

And of course, you'll want to build a culture where people are encouraged to speak up and ask questions—without fear of feeling silly. Leaders can model this in their own behavior, especially if they practice humility. And remember, the language that leaders use matters (see Chapter 15). So rephrase failures as learning opportunities, and remember the power of *yet*. If people don't know how to do something, they just don't know how to do it *yet*.

Key Takeaways

Let's quickly recap the key points on curiosity and continual learning:

- Curiosity is the desire to learn and understand new things. This desire fuels a journey of continual learning (also called lifelong learning), which is the ongoing, self-motivated pursuit of knowledge.

- Two elements in particular are essential to curiosity and continual learning: humility and a growth mindset (that is, a belief that everyone has the capacity to grow, learn, and improve).

- Curiosity is a natural driving force that propels us on to new, interesting experiences. If you can spark your curiosity and keep that spark alive through continual learning, you'll be giving yourself the best chance of a successful, fulfilling life (workwise and otherwise). Curiosity is also an important factor in business performance.

- To boost your curiosity and embrace continual learning, try to cultivate humility and a growth mindset. (Reading Carol Dweck's *Mindset* is a great starting point.) Other good practices are to ask lots of open-ended questions, listen with interest to what others have to say, and seek out new experiences and learning opportunities.

Curiosity and continual learning are obviously helpful when it comes to navigating changes in life and at work—especially if you're willing to continually invest in your skills and knowledge. In the next chapter, we'll explore other ways individuals and organizations can embrace change.

Notes

1. Hal Gregersen: 'Teachers should reward questions, not just answers'; WIRED; https://www.wired.co.uk/article/hal-gregersen#:~:text=According%20to%20a%20paper%20in,average%20four%2Dyear%2Dold

2. The business case for curiosity; Harvard Business Review; https://medium .com/@EngageInfotech/the-business-case-for-curiosity-db133a900bcd

CHAPTER 19
EMBRACING AND CELEBRATING CHANGE

As we saw in Chapter 12, we're facing an acceleration of technology-driven change, to the extent that hundreds of millions of people may need to switch jobs or learn new skills within the next decade. Adaptability gives us the mental resilience to respond to changes—big and small—and learn to roll with the punches (circle back to Chapter 12 for more on the mindset side of change). But what about the practical side of change? How can individuals and businesses learn to embrace, and even celebrate, change? Read on to find out.

What Do We Mean by Embracing and Celebrating Change?

Change has always been a part of life—it's the foundation on which progress and evolution are built (for individuals, organizations, and our species as a whole). Sometimes this change can be scary. We can all identify with this, having lived through a recent pandemic in which (in the early days at least) the situation was rapidly evolving. Change can be a hard thing to move through.

Learning to embrace and celebrate change certainly makes life easier. But what does this entail? Obviously, adaptability plays a huge role in embracing change (again, see Chapter 12), but let's focus on the practical side of things. For me, the practical side of embracing change involves two different processes: learning to manage change successfully (i.e. driving and implementing change) and learning to navigate change from a more personal perspective (i.e. when you're on the receiving end of change). Both are essential—although, to be fair, some people are better at driving change than being on the receiving end!

Whether you're driving or navigating change, you may face various challenges. When you're tasked with making sure a change happens—and this can be anything from moving house to restructuring an organization—there will be many practical steps needed to implement the change, and each step can throw up challenges. One of the biggest roadblocks is resistance to change. The fact is, if people refuse to change, then the change can't happen.

It's not much easier when you're on the receiving end of change. You'll have to overcome your own resistance to change before you can navigate the change itself.

Why is change so hard?

As scientist and MIT lecturer Peter Senge puts it, "People don't resist change. They resist being changed." Why is that the case? Largely it's because of the human tendency towards habits. Think about your day-to-day life and how much of it revolves around set habits: the time you get up and go to bed, what you eat for breakfast, the exercise you do, how you spend your evenings, the hobbies you enjoy, and so on. When something in that routine changes, it takes time to adapt to the new reality (think about changing a password and then repeatedly going to type in the old password for weeks after).

This is because habits easily become ingrained. Then, when change happens, we have to work hard to overcome the ingrained behavior. Eventually the new behavior becomes just as ingrained, but that period in between leaving the old habit behind and getting comfortable with the new status quo can be uncomfortable or frustrating. And if the change is a big one, such as an organizational restructure with potential job losses, there may even be a hefty amount of fear involved. Fear is very often at the heart of resistance to change. But resistance isn't the only obstacle to overcome. Other obstacles can include:

- The sheer scale of the change: A big change can feel so much harder to navigate and implement than a small change.

- Previous attempts at change resulting in failure: This can make us skeptical about other changes that come our way.

- Uncertainty: As *Thinking Fast and Slow* author Daniel Kahneman points out, most of us would rather be wrong than uncertain—we prefer any form of resolution, even a wrong one, over no resolution at all. But when undergoing any change, the outcome is usually uncertain. There's no guarantee of success, and this is uncomfortable.

Change models

Because change is hard, experts have put a lot of time and energy into unlocking the process of change, and developing models to help people understand, manage, and navigate change.

For example, did you know that there's a proper formula for change? Originally proposed by David Gleicher[1] and later revised by Kathleen Dannemiller, the formula is designed to help business leaders assess the likely success of change. It looks like this:

$$C = D \times V \times F > R$$

C is change. D is dissatisfaction with the status quo. V is a vision that things could be better. F is first concrete steps that can be taken towards that vision. And R is resistance.

So, according to the formula, if the product of dissatisfaction, vision, and first concrete steps is greater than resistance, then change is possible. But if any of those three vital factors are missing or low, resistance will win out and the change will fail.

Another model that I really like is the "J-curve" change model (also known as the Satir System), a five-step process developed by family therapist Virginia Satir. The model is designed to help individuals process change, and support others as they process change.

It goes like this:

- Stage 1 is late status quo. Here, the situation is familiar and consistent. People know exactly what to expect, what is expected of them, and how to behave. There's a sense of security.

- In stage 2, a foreign element, driving factor, or threat is introduced that changes the status quo. The sense of security and stability is threatened. The result is very often resistance.

- Stage 3 is (rather alarmingly) called chaos. Here, the people involved are plunged into unfamiliar, unpredictable territory. As a result, they may feel stressed, uncomfortable, confused, or even scared. In a work context, this is when performance drops. In a relationship context, the relationship suffers.

- Stage 4 is integration. Here, people discover the positive side of the foreign element (known as a *transforming idea*). A good example is discovering that implementing a new process actually makes employees' lives easier. This is when the change begins to feel exciting, and people get on board with the change. Performance and relationships improve.

- Finally, we have stage 5, the new status quo. By this point, people feel comfortable with the change. Performance stabilizes at a higher level than before. Eventually, the new status quo becomes a late status quo and the process begins again.

Although this model was developed to describe the patterns of behavior among people undergoing family therapy, it can be applied to any group of people who are experiencing change. The key takeaway is that change has a natural process, and that things generally get worse before they get better—hence, the J-curve shape, where performance or relationships take a dip before climbing to new heights. Later in the chapter we'll explore practical ways to address these five stages.

A super-quick introduction to change management

Now let's move onto implementing change. We know that the process of transitioning from one status quo to a new status quo can be tricky. And this is where change management helps. Change management is a practical framework for, you guessed it, managing change. It's basically a series of tasks that, together, ensure a transition from one situation to a new situation, with as little disruption as possible. It ensures change is managed effectively—whether the change is in response to something unforeseen (such as business disruption), or part of a concerted effort to improve the organization.

Change management involves both the individual element (for example, ensuring the needs of individuals are taken into account), and the organizational element (for example, ensuring any organizational obstacles are overcome). There's also enterprise change management, which is geared towards systematic change across the entire organization. There are lots of change management tools and methodologies you can follow, and I talk more about this later in the chapter.

Why Is It Important to Embrace and Celebrate Change?

Especially in the workplace, change is a continual presence. Whether it's driven by new technology, shifts in the market, expansion, or cutbacks, change is a constant. And if you think you've seen an accelerated pace of change in recent years, you ain't seen nothing yet. The pace of change will continue to accelerate throughout the fourth industrial revolution.

Change is often hard. But we can't deny that change can also be a powerful force for good. If life never changed, we'd end up bored, stuck, uninspired. Think of a habit you successfully changed or a new job that ultimately led to a more fulfilling work life; it may not have been easy at the time, as you adjusted to the new status quo, but it was worth it. This is the J-curve in action.

Yet, despite the potential rewards waiting at the end of the change tunnel, the truth is that most change initiatives fail—a whopping 70 percent to be exact.[2] This may be due to lack of adaptability (see Chapter 12), or it may be because the organization has failed to properly implement the change and help people navigate it. When something in the change process goes wrong, it can have many negative knock-on effects, including:

- Resistance to (or fear of) future change
- Loss of motivation
- A decline in productivity
- An exodus of talent
- A disconnect between the organization's leadership and employees
- The organization being left behind, as competitors adapt more easily

Bottom line, we need to get better at change. We must all learn to embrace and celebrate change as a positive force. This brings us to the next point.

How to Embrace and Celebrate Change

It's impossible to embrace and celebrate change if you lack the practical skills to deal with change. Whether you're the one driving change or the person on the receiving end of change, the following tips will help.

For individuals who are navigating change

If you're on the receiving end of change, I recommend that you:

- First, acknowledge that change is happening. There's no point burying your head in the sand. Acknowledge that change is an inevitable and normal part of life.

- Assess your level of adaptability and look at practical ways to boost your flexibility. (This is the focus of Chapter 12.)

- Gather information. When you're facing change, it can be tempting to leap into action. But that's not always a good move. Instead, take a breath and deploy your critical thinking skills (Chapter 5). Ask questions and gather information to help you assess what exactly is changing, how it's changing, and why it's changing— think back to the change formula and seek to understand the *vision* behind the change and the *first concrete steps* that need to happen.

- Recognize and acknowledge your feelings about change (see Chapter 7, "Emotional Intelligence and Empathy"). But try not to react based on these feelings. It's normal to be wary of change, but you want to respond from a place of considered thought, after gathering the information you need. You may need some time to process the information before you respond.

- Don't assume the worst—think of the best. Remember, this is about embracing and celebrating change. Try to flip the narrative and visualize the best-case scenario that will result from the change. How will your job or life get better?

- Look for ways you can control the change. You may not be in the driver's seat, but there will probably be areas where you can exercise some control. You might, for example, plan out the steps you personally need to take as part of this wider change. You may also be able to plot your own timetable for achieving certain milestones, within the organization's overarching schedule.

- Set yourself some learning goals that will help you navigate this particular change. (See Chapter 18, "Curiosity and Continual Learning.") For example, if you're taking on new responsibilities, what courses or books will help you gain confidence in that role?

- Eat the elephant one bite at a time. The old "How do you eat an elephant?" adage is useful when it comes to navigating change, because it reminds us to not get overwhelmed by the big picture. Focus on the first step, and don't think about the second step until you've completed the first one. And so on and so on.

- Give yourself time. Changing habits and learning new behaviors can be a slow process—in fact, the average time it takes to adopt a new habit is 66 days. Be patient.

- Be kind to yourself when things don't go according to plan. Remember the shape of the J-curve, when performance and relationships typically dip before they get better? Get familiar with the J-curve model and reassure yourself that it's all part of the process.

- Stay connected to your coworkers. Make sure you have ways to communicate with others in your team, especially if you're not all in the same place.

- Related to the previous point, get other people's perspectives. Talk to your coworkers about the change you're experiencing and how it makes you feel. How are they feeling? How are they navigating the change personally? That said, do try to seek out those who consistently show a positive attitude. You know who they are. And you know who to avoid—those inflexible Negative Nellies who see any change in the workplace as a personal affront!

- Bang your own drum. Celebrate the big and little wins whenever you can, even if it's just a mental pat on the back. If you can, celebrate with your team when you achieve common goals.

- Reflect on your resilience. Think about it—with every goal achieved, and every change you navigate, you're ultimately becoming a stronger, wiser, more adaptable person. Isn't that amazing?

Change management tips for organizations and people driving change

There's a vast number of change management processes out there. Go exploring and find a model that works for you. Following a specific change management process, and using tools and templates for various stages (for example, tracking tools) will make your life easier.

One great example of a change management methodology is "The 8-Step Process for Leading Change," developed by Dr. John Kotter (there's an e-book you can download at kotterinc.com). Very briefly, the eight steps are:

1. Create a sense of urgency—by communicating the importance of why the change is needed now.

2. Build a guiding coalition—a team to guide, coordinate, and communicate with the people involved in the change.

3. Form a strategic vision and initiatives—a vision of how the future will be different to the past, with initiatives linked to that vision.

4. Enlist a volunteer army—people who will rally around the common purpose and help drive change.

5. Enable action by removing barriers—such as inefficient company processes that will hamper the change.

6. Generate short-term wins—this helps you track progress and bring more people on board with the change.

7. Sustain acceleration—basically, press harder after the first successful changes.

8. Institute change—make a connection between the new behaviors and organizational success, until the new behaviors become ingrained.

I also encourage you to learn more about the J-curve model of change, particularly how people feel at the various stages (and, in turn, how you can support them). For example:

- At stage 1 (late status quo), you can encourage people to look for new ways of doing things.

- At stage 2 (foreign element and resistance), you can encourage people to share their feelings.

- At stage 3 (chaos), you can support people as they try out new ways of working, and let them know that it's okay to fail or feel frustrated.

- At stage 4 (integration), you can support relationships and dialogue as people begin to adapt to the change.

- At stage 5 (new status quo), you can celebrate success.

So, to successfully manage change, you need a clear change methodology and a good grasp of how change will affect the people within the organization (so you can support them).

Here are some other, more general tips for driving change in your organization:

- Start with quick wins. We know that failed change can result in skepticism of future change. But on the flip side, implementing a successful change creates a positive track record. With this in mind, look for the small or quick wins that will help you build a positive track record of change. (And if the small change doesn't pan out, at least the negative impact of a small change is less than if you'd gone straight for a big change.)

- Remember, people are generally more comfortable with evolution, not revolution.

- Break down goals into smaller milestones. This is essential for accountability and tracking progress, but it also helps with maintaining momentum and enthusiasm.

- Clear communication is key. This is mentioned in the Kotter model, but let me just stress again that you need to communicate your vision, why the change is needed, and how it will impact people in the organization. You will also need to listen to people's concerns. How will you encourage people to speak up, and how will you address their fears?

- Find key influencers. One very potent tactic is to identify those unofficial leaders within the organization that people naturally follow. Bring these people on board with the change at the earliest opportunity and you'll stand a much better chance of bringing the rest of the team along.

- Practice empathy. Put yourself in others' shoes. How would you be feeling about this change if you were them? Revisit the J-curve stages to understand the emotions that may arise at each stage.

- Prepare for problems along the way. There will inevitably be dips in performance or small failures as people adapt to the new way of doing things. Try to anticipate potential sticking points in advance so you can address them before they become a major problem. Potential obstacles might include company policies and procedures, lack of training or tools, and resistance to change.

- Celebrate successes, big and small. Ultimately, you want to build a culture where change is seen as something to celebrate. So highlight and reward success during the change process, and after. Over time, change becomes a positive force.

- Encourage continual improvement. Once the new status quo has been achieved, don't rest on your laurels. Encourage team members to constantly identify new ways of doing things. You want to create a culture of continual, incremental improvements, where change becomes part of the organization's DNA.

- Wherever possible, look at upskilling and reskilling versus job losses. In the fourth industrial revolution, skills will quickly grow stale. Accept this as an inevitable part of your organization's evolution, and seek to build new skills within wherever you can.

Key Takeaways

In this chapter we've learned:

- With 70 percent of change initiatives resulting in failure, it's clear that we all need to get better at change.

- Learning to embrace and celebrate change involves two different processes: first there's managing change successfully (i.e. driving and implementing change), and second there's navigating change from a more personal perspective (i.e. when you're on the receiving end of change).

- There are many practical steps you can take when personally navigating change. For example, you can work on your adaptability (Chapter 12), ask questions, gather information, set yourself goals, focus on one step at a time, and be patient with yourself as you adapt to the new status quo.

- To successfully manage change, you need a clear change methodology and a good grasp of how change will affect the people within the organization (and, in turn, how you can support them).

When driving or navigating change, it's more important than ever to look after yourself and make time for those self-care rituals that help you feel calm, rested, and a little more in control of life. Which brings us neatly onto the final skill in this book.

Notes

1. Formula for Change; Psychology Wiki; https://psychology.fandom.com/wiki/Formula_for_Change

2. Changing change management; McKinsey; https://www.mckinsey.com/featured-insights/leadership/changing-change-management

CHAPTER 20
LOOKING AFTER YOURSELF

With so much going on in the world, and with the rapid pace of change that characterizes the fourth industrial revolution, it can all seem a bit overwhelming. I'm not immune to this. I love what I do and yet I still have times where I feel weighed down by the length of my to-do list, stunned by the pace of change, or just plain worn out. You probably have similar moments. Maybe those "moments" are becoming more common or prolonged amidst the crazy pace of life today. This is precisely why you need to look after yourself, take care of your physical and mental health, and find more balance in life.

Let me state up front that this is an area I'm constantly working on. I can't pretend I've achieved the perfect work–life balance. (Because I love my job, I sometimes struggle to step away and leave work behind.) I can't pretend that I never feel stress or worry. Nor can I pretend that I'm always in tip-top physical and mental shape. Looking after myself is a journey, just as it is for most people. And that journey isn't effortless. Prioritizing self-care, making time to look after myself, being strict about my work boundaries, spending more time with my wife and children, and all those good things requires a conscious effort on my part. So rest assured that I'm not going to preach things I don't do myself. But I will share the tools and techniques that I find helpful.

Of course, not everything in this chapter will resonate with you, so I encourage you to explore other areas and habits that I perhaps haven't mentioned but that interest you personally. (For example, maybe water-color painting is your ideal stress-reliever. Or boxing. Or decluttering your house with Marie Kondo–type rigor.) Feel free to design your own self-care journey, and discover those steps that help you live life with more ease. Think of this chapter as a jumping-off point.

What Does Looking After Yourself Entail?

All these notions of looking after your mental and physical health, lowering stress, having more time for the things that matter, and so on come down to one thing: finding *balance*. There's a common misconception that "work–life balance" means spending equal amounts of time on work and nonwork life (like balancing out two sides of a measuring scale), but that's not true. Balance may mean working four hours a day to you, while others thrive on the structure of a longer workday. Finding balance simply means being able to separate your work and nonwork life, and, crucially, feeling fulfilled in *both* areas.

In general, when I talk about looking after yourself and finding balance, I mean:

- Being able to stay on top of your workload and meet deadlines, without working all hours

- Spending quality time with your children, partner, friends, and other people who matter to you

- Maintaining boundaries between work and nonwork life, so you're not worrying or thinking about work all the time

- Making an effort to eat nourishing foods, exercise regularly, relax, and keep up with the hobbies/activities you love

- Having a proper and restful sleep routine (which may mean nine hours for you, or six hours for someone else)

This is the ideal. You may be a long way from this at the moment, or you may put more effort into some areas than others. That's okay.

Forget perfection

The idea behind looking after yourself and finding balance isn't to pile on further stress and goals, or to make you feel like a failure for *not* doing certain things. Rather, it's about taking small, practical steps towards a more balanced, more content life. It's about understanding what you need to do in order to feel fulfilled in both areas of your life—work and nonwork—and then committing to those steps.

You'll notice that the word "time" crops up a lot in this chapter. But let me stress that balance isn't about packing more into your day or being more productive. That's not to say time management isn't an important skill (it is; see Chapter 17)—but it's not the secret to finding balance. Balance is about feeling fulfilled and content.

Easily said. Not so easily done. In fact, it feels like modern society is setting us up to fail at this whole balance thing. Gone are the days when (most) families could live on one income, which means many of us are juggling a full-time job alongside the responsibilities of caring for a family (and/ or other caring responsibilities). Finding time for things like hobbies or exercise may seem like a pipe dream! However, true "balance" means taking care of yourself as well as others. In order to show up for others, be a good carer or parent, be a good partner, *and* perform well at work, you have to show up for yourself as well.

As part of this, you may need to make some tough decisions. You may need to say no to certain things, assert stronger boundaries at work, have a more disciplined morning routine, or whatever. I'm not saying it's a piece of cake (remember, this is something I have to work at, too). But a more balanced, more content life is certainly worth striving for. And like all journeys, even tiny steps add up over time.

How's your balance at the moment?

Before we move on, it's worth assessing where you are right now in terms of balance. Ask yourself:

- What am I currently prioritizing? Am I paying more attention to one part of life than the others? How are the other areas affected?

- How am I feeling, emotionally and physically? Am I often frustrated, angry, or tired?

- What is causing me to feel that way? Are there, for example, specific things that are causing stress?

- What needs to change? The practical tips at the end of the chapter should help you devise an action plan that works for you.

Introducing the SHED method

When I came across the SHED Method, developed by performance coach Sara Milne Rowe,[1] it immediately struck a chord with me. SHED stands for Sleep, Hydration, Exercise, and Diet—and according to Milne Rowe, building positive routines in these areas helps us to feel more in control of life. But she also talks about tapping into our five key energies, and this is the part of the book that I found particularly interesting.

The five energies are:

- Body energy. The body is the foundation of the other energies, so this is all about doing things that give your body energy. This is the SHED part, basically—taking care of your Sleep (and rest), Hydration, Exercise (which includes movement of any kind), and Diet. By doing this, you fuel your next energy, mood.

- Mood energy. This involves finding ways to keep your mood calm and positive, and learning how to change your mood when it's not serving you (for example, taking some deep breaths when you feel anxious).

- Mind energy. Positive mood energy in turn boosts your mind energy, which gives you the ability to concentrate, solve problems, make better choices, be curious, and so on.

- People energy. Who you spend time with affects how you feel. Therefore, this energy is about surrounding yourself with people who boost your energy—rather than sap it.

- Purpose energy. This means connecting to whatever drives you (usually something outside of your own self-interest) so that you feel excited about life. Without purpose—without the "why am I doing this?"—life can easily feel like a treadmill.

Milne Rowe says that managing these five energies has a huge impact on your ability to be and feel your best. It makes sense when you think about it—making any sort of sustained change in life (in this case, finding more balance and building lasting self-care habits) requires commitment. And commitment requires energy. If you can increase your energy by building good SHED habits, doing things that boost your mood (and mind), spending time with the right people, and connecting to your bigger purpose, it'll be much easier to make lifestyle changes stick.

Why You Need to Look After Yourself

For me, this is a topic that feels particularly timely. Our world is unpredictable and changing fast. There are so many big issues facing society. Technology brings new challenges and, very often, a pressure to be constantly connected. And the lines between home life and work life are becoming increasingly blurred, especially if you work for yourself, or work from home. Now more than ever, we need the skills to look after ourselves, so that we can perform well at work, achieve our personal and professional goals, be there for our loved ones, and feel our best. After all (as far as we know), we only get one bash at life. That in itself is a compelling reason to look after yourself and live a more balanced life—there's no second chance to come back and do it all again.

But let's explore some more earthbound arguments for self-care and balance.

Less stress

Stress can have a negative impact on your body, sleep, emotional state, and even your behavior. It can contribute to health conditions such as high blood pressure, heart disease, and diabetes.[2] And even comparatively mild physical effects—things like chest pain, fatigue, muscle tension, headaches, upset stomach, heartburn, and poor sleep—can take a significant toll on your physical and mental well-being. You may feel depressed, or go off sex. You may catch cold after cold after cold, because your immune system has been weakened. You may experience relationship problems, because you're snappy and distracted. Stress can leach into every facet of your life, making it difficult to get even a moment's peace.

What I find really concerning is the sheer prevalence of stress these days. One UK study found that 74 percent of people have felt so stressed they were overwhelmed or unable to cope (this was pre-pandemic).[3] And when we feel overwhelmed, even small everyday obstacles or decisions can feel insurmountable. The American Psychological Association surveyed Americans about their stress levels during the pandemic and found that a third were so stressed they were struggling to make basic decisions, like what to wear or eat.[4]

Stress can have many causes: ill health, the ill health of a loved one, financial uncertainty, global uncertainty, comparison to other people's (seemingly more successful) lives. But there's no denying that work is one of the biggest causes of stress. According to one survey in the US, a whopping 83 percent of workers suffer from work-related stress.[5]

The thing with stress is that we often turn to coping mechanisms that are exactly the opposite of what we *should* be doing to alleviate the stress. Almost half of people from that UK survey admitted they turned to unhealthy food because of stress, and around a third either starting drinking or drank more. It's a classic catch-22 situation; because we feel stressed, we eat crap, or polish off a bottle of wine, or laze on the sofa instead of going for a walk. Then we feel even worse as a result.

Since we're very unlikely to eliminate all forms of stress, we need to get better at lowering our stress levels. Taking care of ourselves and creating a more balanced life is the way forward, because it involves doing all the things that help us feel calmer, more focused, less prone to overwhelm, and ultimately less stressed. I cover these practical strategies later in the chapter, but we're basically talking about eating better, sleeping better, exercising, having firm work boundaries, and so on.

Better physical and mental health

Aside from lowering your stress levels, there are other physical and mental benefits to creating a more balanced life. Looking at the physical side, the human body isn't designed to spend eight hours a day, or more, indoors hunched over a desk. We need to move, be outside, get fresh air. Part of leading a more balanced life is creating the time in your daily schedule to get away from the desk and move your body, whether it's simply walking around the block on your lunch break, going for a run after work, or indulging in whatever form of movement floats your boat.

Then there are the mental benefits. I know I feel calmer when I'm getting the balance right. And a calmer mind is generally better able to deal with anxiety, challenges, and negative thoughts when they occur (as opposed to a stressed mind, which might react with panic or catastrophize problems into a much bigger deal than they are). In other words, when you feel balanced and in control, you have the mental breathing room to recognize and process thoughts in a healthier, more considered way, as opposed to just reacting to them.

Better relationships

I'm not just talking about having more time for family and friends here, although that is a clear benefit of finding balance. More importantly, balance gives you the ability to be *present* when you're with others. You're not distracted by work emails on your phone. You're not thinking about work. You're not snappy or irritable. You're not tired. Of course, sometimes you will be these things. Nobody's perfect. But by implementing habits that help you find balance, the idea is you'll generally be less irritable, tired, stressed, and so on.

Better performance at work

You know those entrepreneurs and business leaders who boast about only getting 4 hours sleep a night and working 20 hours a day? I mean, good for them, if that truly makes them feel content and fulfilled. But the rest of us should leave them to it. Because success doesn't require a one-track, all-or-nothing mindset. You can thrive at work, and still have plenty of bandwidth to thrive in all the other areas of life. (Again, achieving this may require you to prioritize carefully, delegate, or set firmer boundaries, but we'll talk more about the practical stuff later.)

I know for me, having a proper work–life balance and having clear boundaries between my work and nonwork life makes me *more* successful at work, not less. When I'm at work, I'm more focused. I can concentrate on getting the work done. And I'm more relaxed because I'm taking care of my body, mind, and mood.

More creativity

Things like stress, ill health, lack of sleep, and other symptoms of imbalance all hamper your ability to think creatively because they take up so much mental capacity. This is a problem because creativity is an important part of living a successful, fulfilling life (see Chapter 8). By finding balance—by building some separation between your work and nonwork life, and prioritizing self-care—you make room for creativity. Think of it this way: have you ever been stuck on a problem, then gone for a quick walk and—bam!—the solution pops into your head? This is what happens when you give the brain space to imagine.

How to Better Look After Yourself

I can't emphasize enough that this is about building habits that work for you—that help you find your own version of balance. The following are tips that I've found helpful at various times, but you don't have to adopt all of these. (Besides, it's not a comprehensive list of every technique out there.) I encourage you to seek out your own self-care habits and routines. A good resource for this is the SHED Method book mentioned earlier in the chapter. Do give it a read and explore your own ways to look after your five energies (body, mood, mind, people, and purpose).

Looking after your body

Starting with physical wellness:

- People who do regular physical activity have a lower risk of all sorts of diseases, from bowel cancer and heart disease to diabetes and dementia.[6] Therefore, it's vital you make time for regular physical activity (walking, running, yoga, etc.). In the UK, the NHS recommends adults do at least 150 minutes of physical activity per week—that's around 20 minutes a day. The easiest approach is to make activity part of your everyday routine, for example, by cycling to work, or walking (rather than driving) to the train station.

- Get outside as often as you can, even if it's just a quick coffee in the sunshine on your morning break.

- Eat a balanced diet, as much as possible made up of whole foods (as opposed to highly processed foods).

- Limit your alcohol consumption.

- Drink more water!

- Practice a good sleep routine. We don't all need nine hours of perfect sleep a night, but we do need a solid, restful sleep routine. So try to go to bed and get up at roughly the same time each day (even on non-workdays). Stay away from your phone, tablet, and other screens in the hour or two before bed. Ensure your bedroom is set up for good sleep (for example, with blackout curtains). And if you choose to have your phone in the bedroom (although, really, you'd be better off just buying an alarm clock), be sure to put your phone into "bedtime" or "do not disturb" mode so you aren't bothered by notifications.

Looking after your mood and mind

A healthy body certainly feeds into a healthy mood and mind, but here are some other steps that have helped me boost my mental health:

- Set strong boundaries between work and nonwork. When I'm not in my office, I'm not working. I'm not checking emails. I'm present with my family or enjoying some quiet time to myself.

- Find ways to unplug and relax. Personally, I find that reading, mindfulness, and running help me to unwind. For you, it could be deep breathing, meditation, long baths, walking in the woods, or getting a massage.

- Make time for hobbies, such as gardening, painting, watching movies, cooking, dancing, or whatever.

- Try to live in the moment, rather than always thinking about what has been and what might happen in the future. Mindfulness is brilliant for learning to be more present because it teaches you to kindly acknowledge your thoughts and feelings, and notice what's going on around you.

- Talk about your feelings, especially when you feel stressed or troubled. And don't be afraid to ask for help when you need it, from friends and family, or local/national mental health services.

- Maintain connections with others. There's nothing like a face-to-face catch-up with a friend or loved one, but when getting together in person isn't possible, make time for a call, video chat, or at least a message.

- Reframe unhelpful thoughts. How we think and how we feel are closely connected. Try to recognize unhelpful, negative thoughts that don't serve you and literally turn them upside down. For example, "This presentation is going to be a disaster" becomes "This presentation is going to go smoothly."

- Let go of worry. It's easy to get into a toxic spiral of anxiety—many of us feel that if we worry about something enough, we can stop it from happening. But your thoughts don't control external events, and if you spend precious time and energy worrying about something before it's happened, you put yourself through the ordeal twice (and that's if the bad thing even comes to pass; often it doesn't). Try to let go and accept that, even if the worst does happen, you have the tools, strength, and support system in place to deal with it.

- Write a letter to your future self. Deep down, you probably know what makes you feel mentally strong, and what to do to take care of your mental health. So, when you're in a good place and feeling strong, write a letter to your future self, to read when things aren't going so swimmingly. In the letter, outline some steps that you know make you feel more positive, and perhaps even list some things that you're grateful for. Save the letter so you can read it in tougher times.

- Embrace imperfection. Perfectionism can lead to stress, so remind yourself that you don't need to be perfect. You don't need to be the best. Being good enough is, well, good enough!

- Remember that finding balance is a journey, not something you achieve once and then you're done. You will constantly have to work at building and maintaining these good habits. It's worth it.

Looking at people and purpose

In terms of these final two energies:

- Look carefully at the people you spend your time with and ask yourself, does this person give me energy or sap my energy?

- Keep company with people who make you feel good—people who are positive and supportive of you.

- Wherever possible, avoid or limit spending time with "energy vampires." You know the kinds of people I mean—people who leave you feeling drained. People who take rather than give. People whose negativity is contagious, so you find yourself echoing their negative speech when you're in their presence.

- Obviously, if you're part of a team, you may not have a say in who's on that team. Even so, you can spend less time listening to the people who sap your energy and more time around the people who infect you with their positive, can-do attitude.

- Purpose is about doing something meaningful with your time and talent. I know this is often easier said than done (we all have bills to pay). But if you can do a job you love, that's a good starting point.

- If your work life doesn't feed your purpose energy, it's not the end of the world—seek out nonwork activities that give you that precious sense of purpose, whether it's volunteering for a local charity, mentoring or educating others, writing a novel, or whatever.

Making a conscious decision to stick to it

As I've said, building a more balanced life may require you to make some difficult choices and reprioritize. Here are some tips for making better decisions that support your journey:

- Say no to things that aren't a priority *for you*. Yes, it's good to help people whenever you can, but there will be times when their request clashes with your own priorities. In that case, learn to kindly but firmly say no (or "no, not now").

- Let go of non-priorities, and outsource or delegate these tasks whenever you can. Circle back to the time management chapter (Chapter 17) for tips on prioritizing your to-do list.

- Eliminate the time-wasting aspects of your day, such as social media notifications that suck you into your phone, and before you know it half an hour has gone by. Switch off app notifications, and ensure your social media activity (or news scrolling, or whatever) takes place on your terms, at times that suit you.

- Treat your time like the precious asset it is. For example, if you're invited to a meeting but you're not sure it's the best use of your time, try saying something like "I don't think I'm the right person to attend this meeting," or "I don't think I can add value to this conversation, but I look forward to seeing the email recap afterwards."

- Set strong boundaries—for example, not answering work emails after 5 p.m. or on the weekend—and enforce those boundaries when people try to test them.

- If you work from home, try to create a dedicated workspace that you go to during work hours, and leave behind at the end of the working day, even if it's just a corner of the spare room or an alcove under the stairs.

- Remember to work smart, not work long (again, see time management, Chapter 17).

- When work expectations are simply too much, speak up. Talk to your manager or HR and let them know that the demands are unsustainable.

A quick word for organizations and leaders

Finding balance isn't just the responsibility of individuals; we need workplaces to step up and help people live more balanced, more content lives. The link between stress and productivity (or rather, lack of productivity) is argument enough for this; numerous studies have identified that greater stress leads to lower productivity.[7]

Bottom line, organizations cannot afford to ignore this topic. That's why I believe businesses must:

- Encourage a culture of openness, so people feel free to speak up when the pressure gets too much.

- Ensure leaders model a good work–life balance, for example, by taking regular breaks, getting out of the office at lunchtime, not emailing people outside of work hours, and so on.

- Offer flexible and remote working where possible.

- Train managers to spot signs of stress and poor work–life balance. Have systems in place to support those who need it.

- Let employees take time off for volunteering activities.

- Encourage activities that promote physical activity and lower stress, for example, through subsidized exercise classes, on-site yoga sessions, or gym discounts.

- Ask your employees what they would like the company to do to boost work–life balance.

Key Takeaways

To recap the key points on finding balance:

- In this age of rapid change and information overload, it's essential to look after yourself, take care of your physical and mental health, and find more balance in life.

- Rather than spending equal amounts of time on work and non-work, true "work–life balance" means being able to separate your work and nonwork life, and, crucially, to feel fulfilled in *both* areas.

- In practice, this looks like this: staying on top of your workload without working all hours; spending quality time with loved ones; maintaining boundaries between work and nonwork; eating well; exercising regularly; making time for hobbies and relaxation; and having a proper sleep routine.

- The SHED Method by Sara Milne Rowe provides a useful blueprint for looking after yourself. (SHED stands for Sleep, Hydration, Exercise, and Diet.) In her book, Milne Rowe talks about the need to look after your five "energies": the body energy, mood energy, mind energy, people energy, and purpose energy.

- Good ways to boost your different energies include building exercise or movement into your everyday routine; practicing mindfulness and meditation; letting go of perfection; surrounding yourself with positive people; and doing things that give you a sense of purpose. Do devise your own techniques that work best for you.

- And remember, looking after yourself is an ongoing journey, not something you achieve once then forget about. Yes, it requires effort, commitment, and discipline, but it's worth it in order to gain a more balanced, more content life—a life that's rich and fulfilling.

This brings us to the end of our voyage through 20 essential skills for success. Now, let's round off with some final thoughts.

Notes

1. The SHED Method: The New Mind Management Technique for Achieving Confidence, Calm and Success; Penguin; https://www.penguin.co.uk/books/294/294581/the-shed-method/9781405941327.html

2. Stress symptoms: Effects on your body and behavior; Mayo Clinic; https://www.mayoclinic.org/healthy-lifestyle/stress-management/in-depth/stress-symptoms/art-20050987

3. Mental health statistics: stress; Mental Health Foundation; https://www.mentalhealth.org.uk/statistics/mental-health-statistics-stress

4. Stress and decision-making during the pandemic: American Psychological Association; https://www.apa.org/news/press/releases/stress/2021/october-decision-making

5. 42 Worrying Workplace Stress Statistics; Stress.org; https://www.stress.org/42-worrying-workplace-stress-statistics

6. Benefits of exercise; NHS; https://www.nhs.uk/live-well/exercise/exercise-health-benefits/

7. Workplace Stress and Productivity: A Cross-Sectional Study; Kansas Journal of Medicine; https://www.ncbi.nlm.nih.gov/pmc/articles/PMC7889069/

FINAL WORDS

In this book, we've learned that understanding the impact of new technologies and being able to work confidently alongside these technologies is certainly an important part of success in the fourth industrial revolution. But over and above that, distinctly human skills like creativity and emotional intelligence will be more valuable than ever before.

What These 20 Skills Tell Us About the Future

When I look back over the chapters in this book, a few common themes jump out at me:

- Humility, in terms of recognizing our strengths and weaknesses so that we can grow and improve

- Optimism, that the ability to surf the wave of transformation is well within our grasp, and nobody need be left behind

- Self-confidence, in that anyone—literally anyone—can learn and improve in these skills

- Resilience, in knowing that these skills can help us successfully navigate whatever changes and challenges may be coming our way

- Taking the initiative, because many of these skills are ignored by traditional education institutions, so it's up to us to curate our own learning path

Just looking at these themes has me feeling optimistic about the future. So when people try to tell you that there's no future for humanity, that machines will take all our jobs, ask yourself: is that the vision of the future you get from this book?

Overall, I'd say the 20 essential skills point to a world in which work is more human, more meaningful, and more fulfilling. Yes, many jobs will evolve as new technologies promise to streamline and automate routine processes. And it may mean that your job title in 15 years' time hasn't been invented yet. The uncertainty that comes with all this change can be challenging, but isn't this vision of the future exciting—a future in which organizations value our very humanness, instead of expecting people to behave like machines? I'm in!

Where to Go from Here

Building the skills in this book will help you thrive in the fourth industrial revolution, and give you the confidence to successfully ride the wave of change. But where should you start?

I say start small. Trying to boost your knowledge of 20 skills at once will only lead to overwhelm. So identify the one, two, or three skills that matter most to you or are the biggest priority for your line of work. Make that your focus for this year. In fact, why not write a letter to yourself outlining the skills that you've prioritized and your learning action plan for the year ahead? (This might include reading books, signing up for online courses, finding a mentor, and practicing the tips and techniques I've highlighted throughout the book.)

Then, when the year is up and you've reviewed your progress, you can repeat the process with the next set of priorities. Build gradually. And do keep in mind that *all* of the skills in this book can be learned. So don't be tempted to dismiss any of these skills because "that's not my thing" or "I'm no good at that." Remember the power of "yet"—as in, "that's not my thing *yet*, but it will be."

Embracing the Road Ahead

Before we end, let's briefly cast our mind back to the previous industrial revolution, which began in the late 20th century and was driven by computerization. Computerization made work and life easier, created value-adding new jobs, and ultimately made the world a smaller, more connected place. (If your memory doesn't stretch back that far, you'll have to take my word for it.) Was it wholeheartedly embraced by everyone at the time? Of course not. But did it ultimately lead to better, easier lives for most people? Absolutely. This fourth industrial revolution will again make the world better and improve our lives—despite the many challenges that come with technology—by freeing up humans to focus our time and talents where they matter most.

For me, these 20 skills point to a future where we celebrate the amazing potential of humans—with all our creative, empathetic, interpersonal talents—to literally create the future we want, and tackle some of the biggest problems facing our world. If we harness our human talents, anything is possible.

Online Bonus Chapter

I have written a bonus chapter to expand on the five human traits that are common themes running throughout this book. These traits underpin each of the 20 future skills:

- Optimism
- Humility
- Self-belief
- Resilience
- Taking the initiative

If you would like some more tips on how to improve these, you can access the bonus chapter at bernardmarr.com/bonus.

In return, I would really appreciate if you could write an online review of the book.

ABOUT THE AUTHOR

Bernard Marr is a world-renowned futurist, influencer, and thought leader in the field of business and technology. He is the author of 21 best-selling books, writes a regular column for *Forbes*, and advises and coaches many of the world's best-known organizations. He has over 2 million social media followers, 1.2 million newsletter subscribers, and was ranked as one of the top 5 influencers in the world by LinkedIn.

Bernard helps organizations and their management teams prepare for future trends and create the strategies to succeed. He has worked with or advised many of the world's best-known organizations, including Amazon, Microsoft, Google, Dell, IBM, Walmart, Shell, Cisco, HSBC, Toyota, Nokia, Vodafone, T-Mobile, the NHS, Walgreens Boots Alliance, the Home Office, the Ministry of Defence, NATO, and the United Nations, among many others.

Connect with Bernard on LinkedIn, Twitter (@bernardmarr), Facebook, Instagram, and YouTube to take part in an ongoing conversation, subscribe to Bernard's podcast, and head to www.bernardmarr.com for more information and hundreds of free articles, white papers, and e-books.

If you would like to talk to Bernard about any advisory work, speaking engagements, or influencer services, please contact him via email at hello@bernardmarr.com

Other books Wiley books by Bernard Marr include:

- *Business Trends in Practice: The 25+ Trends That Are Redefining Organizations (Winner: Business Book of the Year 2022)*

- *Extended Reality in Practice: 100+ Amazing Ways Virtual, Augmented and Mixed Reality Are Changing Business and Society (Specialist Book Category Winner: Business Book of the Year 2022)*

- *Tech Trends in Practice: The 25 Technologies That Are Driving the 4th Industrial Revolution*

- *Artificial Intelligence in Practice: How 50 Successful Companies Used AI and Machine Learning to Solve Problems*

INDEX

INDEX